T0139140

Applied Learning Algorithms for Intelligent IoT

Applied Learning Algorithms for Intelligent IoT

Edited by
Pethuru Raj Chelliah
Usha Sakthivel
Susila Nagarajan

CRC Press is an imprint of the
Taylor & Francis Group, an **informa** business
AN AUERBACH BOOK

First Edition published [2022]
by CRC Press
6000 Broken Sound Parkway NW, Suite 300, Boca Raton, FL 33487-2742

and by CRC Press
2 Park Square, Milton Park, Abingdon, Oxon, OX14 4RN

CRC Press is an imprint of Taylor & Francis Group, LLC

ISBN: 978-0-367-63594-7 (hbk)
ISBN: 978-1-032-11321-0 (pbk)
ISBN: 978-1-003-11983-8 (ebk)

DOI: 10.1201/9781003119838

Typeset in Caslon
by MPS Limited, Dehradun

Contents

Contributors

J. V. Thomas Abraham
School of Computer Science and Engineering (SCOPE)
Vellore Institute of Technology
Chennai, Tamil Nadu, India

D. Aishwarya
SRM Institute of Science and Technology
Kattankulathur, Tamil Nadu, India

Sruthi Anand
Department of Information Technology
Sri Krishna College of Engineering and Technology
Coimbatore, India

V. Anandkumar
Department of Information Technology
Sri Krishna College of Engineering and Technology
Coimbatore, India

Mohan Aparna
School of Computer Science and Engineering (SCOPE)
Vellore Institute of Technology
Chennai, Tamil Nadu, India

Pethuru Raj Chelliah
RJIL College of Engineering and Technology
Bengaluru, India

Esther Daniel
Department of Computer Science and Engineering, Technology
Karunya Institute of Technology and Sciences
Coimbatore, India

S. Durga
Department of Information Technology
Karunya Institute of Technology and Sciences
Coimbatore, India

Jereon Hak
M.S. Student
Rochester Institute of Technology
Rochester, New York, USA

A. P. Jyothi
Department of Computer Science and Engineering VTU
Chennai, India

T. R. Kalaiarasan
Department of Information Technology
Sri Krishna College of Engineering and Technology
Coimbatore, India

K. Karunamurthy
Vellore Institute of Technology
Chennai, Tamil Nadu, India

G. Jaspher W. Kathrine
Karunya Institute of Technology and Sciences
Coimbatore, India

M. V. Kaviselvan
Department of Mechanical Engineering
Sri Sairam Engineering College
Chennai, Tamil Nadu, India

Gotluru Arun Kumar
School of Computer Science and Engineering (SCOPE)
Vellore Institute of Technology
Chennai, Tamil Nadu, India

M. Suresh Kumar
Department of Information, Technology
Sri Sairam Engineering College
Chennai, Tamil Nadu, India

R. Maheswari
School of Computer Science and Engineering (SCOPE)
Vellore Institute of Technology
Chennai, Tamil Nadu, India

Ashwini R. Malipatil
Department of CSE
Sri Krishna College of Engineering and Technology
Coimbatore, India

R. I. Minu
SRM Institute of Science and Technology
Kattankulathur, Tamil Nadu, India

Anju S. Pillai
Amrita Vishwa Vidyapeetham
Coimbatore, India

J. Pushpa
Jain University
Bengaluru, India

S. Rachel
Department of Information Technology
Sri Sairam Engineering College
Chennai, Tamil Nadu, India

Priscilla Rajadurai
St. Joseph's Institute of Technology
Chennai, India

A. M. Ratheeshkumar
Department of Information Technology
Sri Krishna College of Engineering and Technology
Coimbatore, India

Usha Sakthivel
Dept. of Research and Innovation
Rajeswari College of Engineering
Bengaluru, India
and
Dept. of Computer Science & Engineering
Visvesvaraya Technological University
Belagavi, Karnataka, India

Vijayalakshmi Saravanan
Faculty Member
Rochester Institute of Technology
Rochester, New York, USA

T. Sheela
Department of Information Technology
Sri Sairam Engineering College
Chennai, Tamil Nadu, India

Ishpreet Singh
M.S. Student
Rochester Institute of Technology
Rochester, New York, USA

Neha Singhal
Department of ISE
Sri Krishna College of Engineering and Technology
Coimbatore, India

T. Subha
Sri Sairam Engineering College
Chennai, India

G. Suganya
Vellore Institute of Technology
Chennai, Tamil Nadu, India

N. Susila
Department of Information Technology
Sri Krishna College of Engineering and Technology
Coimbatore, India

Emanuel Szarek
M.S. Student
Rochester Institute of Technology
Rochester, New York, USA

M. Umadevi
Department of Computer Science and Engineering
SRM Institute of Science and Technology
Chennai, Tamil Nadu, India

R. Valarmathi
Department of Computer Science and Engineering
Sri Sairam Engineering College
Chennai, Tamil Nadu, India

Pulugu Yamini
School of Computer Science and Engineering (SCOPE)
Vellore Institute of Technology
Chennai, Tamil Nadu, India

1

Convolutional Neural Network in Computer Vision

D. AISHWARYA[1] AND R.I. MINU[2]

[1]*Research Scholar, SRM Institute of Science and Technology, Kattankulathur, Tamil Nadu, India*
[2]*SRM Institute of Science and Technology, Kattankulathur, Tamil Nadu, India*

Contents

DOI: 10.1201/9781003119838-1

Introduction

Emergence of artificial intelligence (AI) has made machines "Smart." By "Smart," we mean the ability of the machine (possibly a computer) to process the given data with rationality, like an intelligent human being, and take a suitable decision. AI achieves this intelligence with the help of its subsets of algorithms (see Figure 1.1).

AI is a broader field that includes machine learning (ML), artificial neural network (ANN), and deep learning (DL) as its subsets. AI is achieved by ANN, which is a network of artificial neurons resembling a biological neuron and the nervous system. ML enables the machines to process the data, make decision, and optimize the results based on experience. ML primarily requires a feature engineer to do the feature engineering process. Also, ML works its best for structured dataset (mostly comprises numerical data) but is proven inefficient for unstructured types.

To process the unstructured and continuous data like video, speech, etc., we need more complex neural network. Here comes in the DL with better results on processing the unstructured data. One

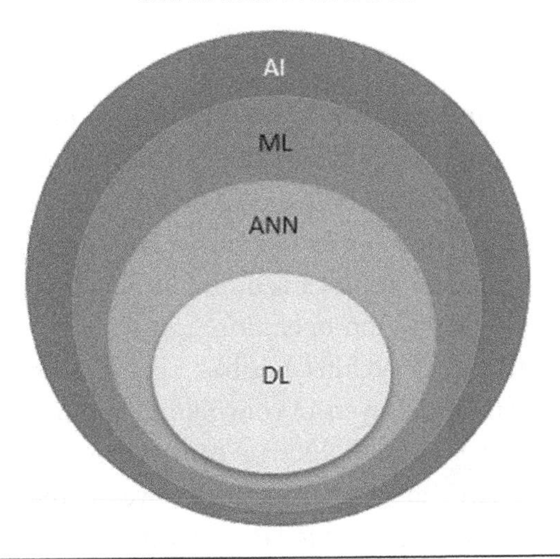

Figure 1.1 Subsets of Artificial Intelligence

of the biggest advantages of DL is that it features engineering automatically, that is, no manual feature selection is required. Also, ML requires manual intervention when the algorithm is struck with optimization problem, which is also automated in DL. DL plays the key role in implementing and visualizing the actual AI. Deep learning, however, requires a large amount of dataset for the purpose of training, which in turn requires a deep network of layers of neurons to process the features and hence the name "deep learning."

DL comprises many algorithms for different types of data. Of those, convolutional neural network (CNN) is mainly used to process images and solve computer vision problems. The idea of CNNs can be traced back to 1980, when K. Fakushima proposed "neocognition" – a hierarchical multi-layered neural networks, which is self-organized and unaffected by the translation operation in the dataset (shift in positions of patterns). It was primarily created for pattern recognition.

The first standard CONVNet was introduced by Yann LeCunn in 1998, in his research article titled "Object Recognition with Gradient Based Learning." It is important to note that Yann LeCunn has cited Fakushima's article in his research work, and the CONVNet of LeCunn has a structure similar to neocognition with back propagation added to the network. Thenceforth, multiple improvements and the extension were made to it.

Convolutional Neural Network (CNN)

Convolutional neural network is primarily used for operations like image classification, segmentation, and other computer vision operations. CNN is also called shape/space invariant artificial neural network (SIAAN) based on its shared weights and translation invariance properties. The key idea behind CNN is to capture the local patterns through convolution operation.

In general, each CNN architecture has a common working flow. A CNN consists of multiple layers of convolution and pooling and a fully connected ANN. A layer in CNN may comprise one convolution layer and one pooling layer together as a single layer if the count of convolution layer and pooling layer is same. Most commonly used type is two convolution layers followed by a single pooling layer.

The number of layers in a CNN architecture depends on the complexity of the application. The output of the last layer is passed to an ANN. The convolution layers and the pooling layers always perform the same operation of learning the input image. The fully connected networks perform the required operations, such as regression, classification, etc.

At each convolution layer, a local pattern is obtained by applying filters throughout the image. Filters can also be mentioned as kernels. The Initial/lower-level layers of the CNN obtain the elementary pattern of the image like edges and lines. The higher-level layers of CNN acquire the patterns recognized by the previous layer, and through the concept of spatial locality, higher-levels obtain larger patterns, as shown in Figure 1.2. The last layer of CNN would

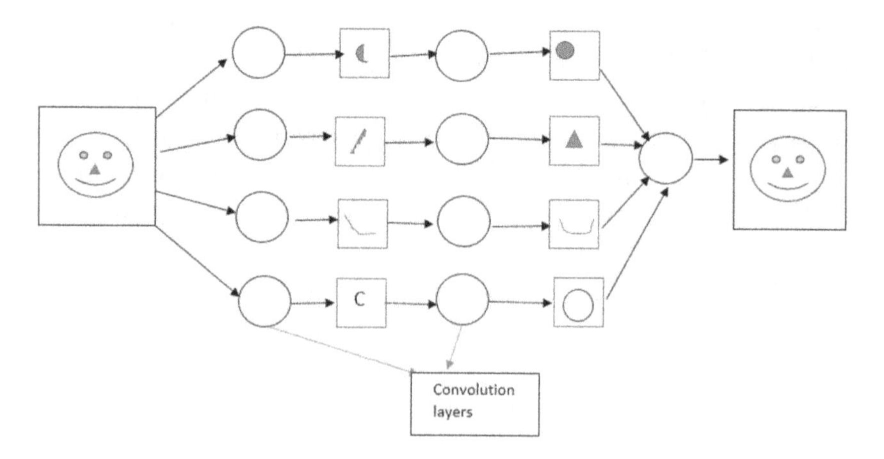

Figure 1.2 Typical Operation of Conv Layer

generate the complete image, and the ANN attached at last would give the required output depending on the problem for which the particular neural network was developed.

Distinctive Properties of CNN

The following properties of CNN make it unique from the other DL algorithms:

- A CNN contains volume of neurons as shown in Figure 1.3, i.e., the neuron in CNN layers are arranged in three dimensions (depth, height, and width).
- Shared Weights: When a filter is applied over a part of an image, same set of parameter values of filter is used for all parts of the image.
- Receptive Fields: Each neuron in the 3-D structure of a CNN is subjected to a defined region of the input data, and this allows the neurons to learn the elementary patterns of the input. This region of input, associated to a particular neuron, is called the local receptive field as in Figure 1.4.

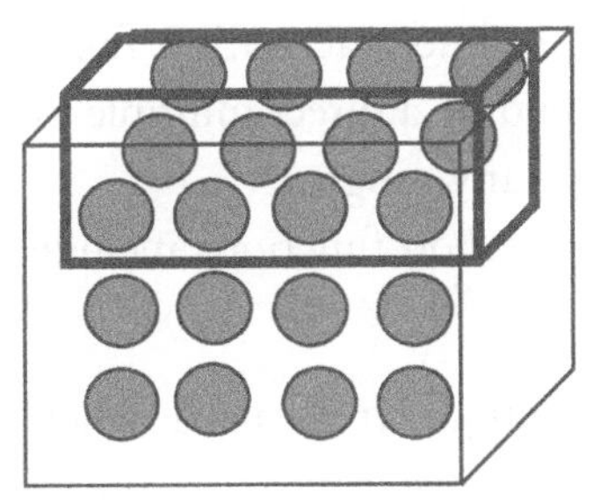

Figure 1.3 Volume of Neurons

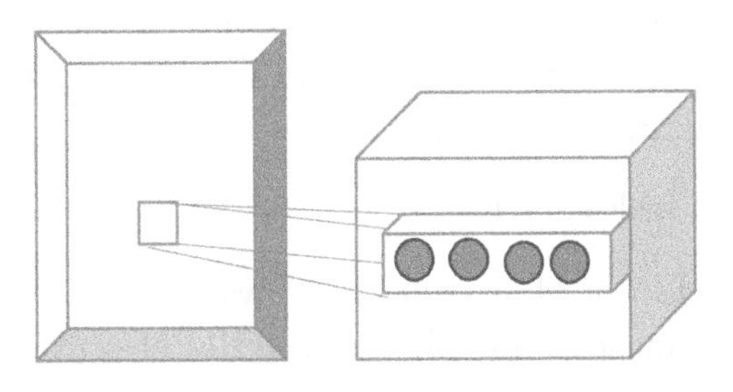

Figure 1.4 Local Receptive Field

- Local Connectivity: The local connectivity pattern between the neurons is essential to exploit spatial locality. This local connectivity pattern enables the network to create the representation of local pattern and then assembles the representation of larger area.
- Pooling: Pooling layer is used to reduce the size of the feature map. A feature map is the output obtained from convolution operation.

Before describing CNN in detail, it is important to know few other terminologies and concepts that would be applied to CNN.

Activation Functions for CNN

An activation function is applied to the output of the convolution layer of the CNN architecture to decide if the output is precise and if it should be passed as an input to the next layer or should it be dismissed.

$$Y = Activation(\sum (weight * input) + bias)$$

The resulting value after the application of activation function ranges usually from 0 to 1 or from −1 to 1, depending upon the function applied. The activation function may either be linear or non-linear.

For any neural network or DL algorithm, linear activation functions are hardly used, since their output would be a direct multiple of the input and the resulting value may be of any range.

The most commonly used non-linear activation functions include the following:

- Sigmoid or Logistic Activation Function: It is used when probability of output is required, since the resulting value ranges from 0 to 1 (see Figure 1.5).
- Tanh or Hyperbolic Tangent Activation Function: The range of the hyperbolic tanh function is from −1 to 1. It is often used for classification between two classes.

Tanh is shown in Figure 1.6.

- Rectified Linear Unit (ReLU) Activation Function: CNN architecture most commonly uses the ReLU activation function. ReLU allows the neural network to converge quite easily and

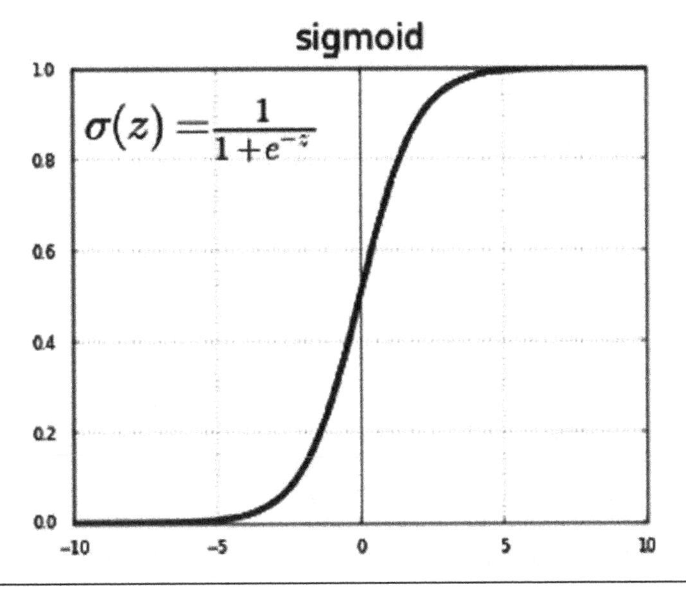

Figure 1.5 Sigmoid Function

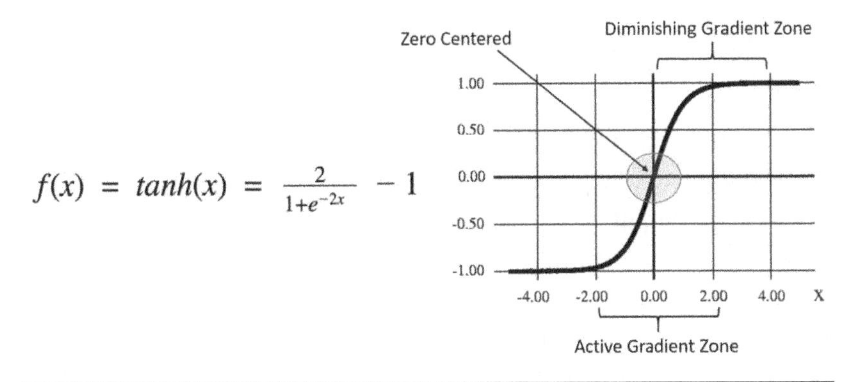

Figure 1.6 Tanh Activation Function

allow back propagation as well. One problem with ReLU is that the gradient of the function becomes 0, when the input becomes 0 or negative. This condition prevents back propagation and is known as dying ReLU problem.

- Leaky ReLU: Also the parametric ReLU (PReLU) prevents dying ReLU problem by introducing a small slope to keep the updates alive.

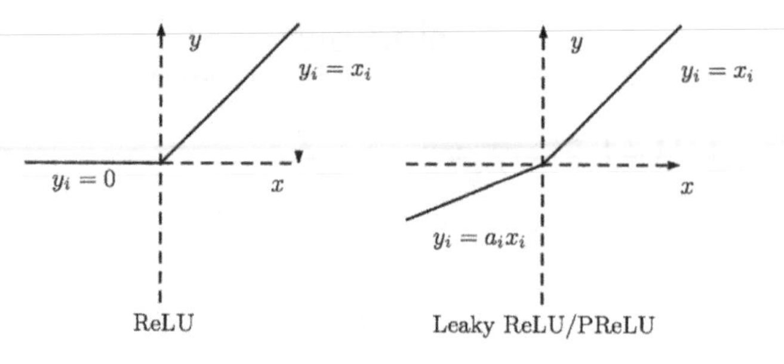

Figure 1.7 ReLU and Leaky ReLU Functions

As shown in Figure 1.7, ReLU works as

$$R(x) = (0, x)$$

But, PReLU/Leaky ReLU function adds the slope value of *a* to the input

$$R(x) = \begin{cases} ax & if \quad x < 0 \\ x & if \quad x \geq 0 \end{cases}$$

If $a = 0.1$, it is called Leaky ReLU and if *a* is any suitable value, it is called PReLU.

- Softmax Activation Function: The most commonly used activation function for multiclass classification is Softmax. It is used only for the output layer unlike the other types of activation function.

Loss Function

The optimization algorithm requires the error value of the current iteration to be estimated repeatedly. This loss function also called error function can be used to estimate the error value of the model thereby, enabling it update the weights of the parameters, so that the accuracy can be improved and the error be decreased in the next iteration. Loss functions are selected based on the problem for which the network is modeled.

- Regression Problem – Loss Functions used are
  1. Mean Squared Logarithmic Error Loss (MSLE)
  2. Mean Squared Error Loss (MSE)
  3. Mean Absolute Error Loss (MAE)

- Binary Classification – Loss Functions used are
  1. Binary Cross-Entropy
  2. Squared Hinge Loss
  3. Hinge Loss
- Multiclass Classification – Loss Functions used are
  1. Multiclass Cross-Entropy Loss
  2. Kullback–Leibler Divergence Loss
  3. Sparse Multiclass Cross-Entropy Loss
- Segmentation – Loss Functions used are
  1. Weighted Cross Entropy
  2. Focal Loss
  3. Dice Loss
  4. Boundary Loss

Datasets and Errors

The dataset for a neural network is split into three different subsets:

Training Data: It is a set of data, comprising 80% of total data, used for learning and modeling the neural network. Training data is used to fit the parameters of the classifier. The error (based on accuracy) arises when a trained model is run over the same training data.
Validation Data: It is a subset of training data held unused, to estimate the accuracy of the model trained using training data based on which the parameters are tuned. The error that arises when the trained model is run over the held out validation data is known as validation error. It provides an estimate of test error.
Test Data: It is the remaining 20% of total data, used to assess the performance of the fully trained neural network. The error that arises when the trained model is run over the test data is known as test error.

Bias and Variance

Bias is an error arising due to erroneous assumption of the model. Higher value of bias causes larger value of training error.

Example: The model might wrongly assume the features to error and miss the essential details of the data while modeling.

Variance is also an error that occurs when the model begins to consider noise as an essential feature and models it. High Variance causes high validation error.

Total Error: The sum of the bias error and variance error is the total error.

Trade-Off between Bias and Variance: For a model to work perfectly, low-bias and low-variance errors are the ideal conditions.

Overfitting and Underfitting

Overfitting occurs when the model trains on the training data very well. Overfitting causes the model to make wrong predictions. False positives increase due to overfitting. High variance and low bias result in overfitting, also known as generalization error.

Underfitting occurs when the model generates accurate predictions on the training set. Underfitting causes false negatives to increase. High variance and high bias result in under fitting.

Appropriate Fit or the statistical fit is the ideal state between overfitting and underfitting (Figure 1.8).

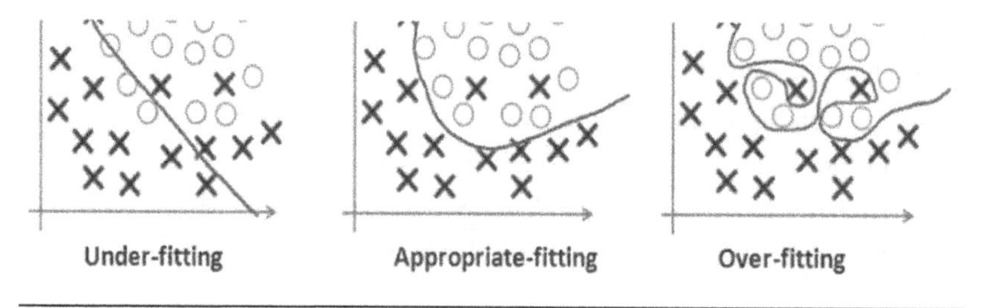

Figure 1.8 Types of Fit

Understanding Padding and Stride

The two main parameters that decide the behavior of the CNN are padding and stride. Both these parameters have impact on the size of the data.

Padding

While processing the images in the convolutions layer, there are possibilities that the features on the edges are lost. Also called the pixel loss at the edges, it might hold the key features required from the images. For the first layer of the convolution, this loss might seem to be less, but for a deeper CNN, the loss is significant and affects the prediction accuracy of the network. To prevent this, the sides of the images can be padded with the empty pixels, i.e., with 0 intensity value, as shown in Figure 1.9. The optimal number of layers of padding "*P*" is calculated using

$$P = \frac{F-1}{2},$$

where, F is the size of the filter applied to the image. If the resulting value of the above formula is in fraction, the division quotient is rounded off to the next higher value. When we apply the equation, the padding type is called same padding, since this helps in keeping the size of the output image after convolution operation same as that of the input image, which usually reduces by the effect of the stride. There is another type of padding, known as valid padding. Valid padding means no padding, i.e., no padding will be done. In valid padding, the part of the image that does not fit the filter would be dropped off.

1	2	3	5	4
6	8	7	9	5
2	6	9	8	6
5	9	8	5	1
5	8	9	9	9

Input image 5X5

After Padding P=1, with F=3, P = (3-1)/2.

0	0	0	0	0	0	0
0	1	2	3	5	4	0
0	6	8	7	9	5	0
0	2	6	9	8	6	0
0	5	9	8	5	1	0
0	5	8	9	9	9	0
0	0	0	0	0	0	0

Input image size 7X7

Figure 1.9 Padding

Stride

Stride is an important component of convolution layer. It denotes how many pixels are shifted by the filter, while it moves over the input image during the feature extraction process (see Figure 1.10). This process can also be stated as convolving. The value of stride depends on the connection of the convolution layer and the pooling layer, the filter size and the properties of images in data. Usually the value of stride is 1. If the values of stride are higher (i.e., more than 1), the dimension of the output is reduced, leading to lesser computation, but there could be loss of feature representation. The value of stride must be an integer. The value of stride plays a key role in trade-off between the accuracy and the running time of the neural network.

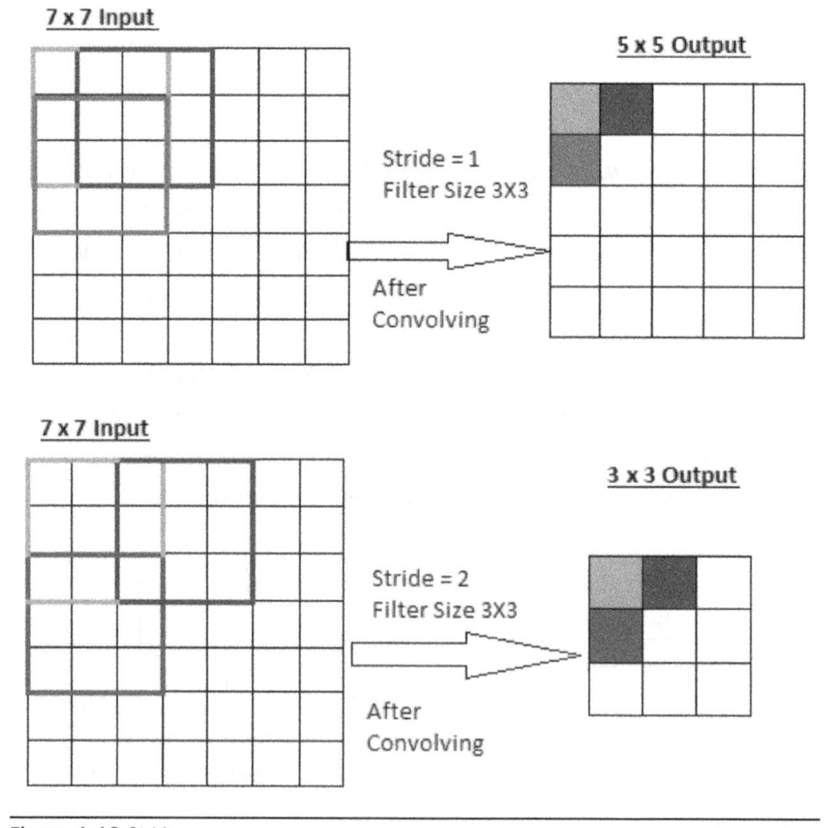

Figure 1.10 Stride

Parameters and Hyper Parameters

Model parameter is a variable whose value is set automatically and is estimated from the training dataset. The values are learned by fitting the model to the data, i.e., they are estimated by the optimization algorithm from the data. These parameter values are used by the model to make predictions. The parameters are different for different applications.

Example: Weights in neural network intercept value in linear regression.

Hyper parameters are set manually before the initiation of the leaning process. These hyper parameters are required for estimating the model parameters. Hyper parameters play a vital role in ensuring good accuracy from the neural network since it controls the learning process of the model. Therefore, it is very important to select an ideal set of hyper parameters and initiate them to optimal values. This process is called hyper parameter tuning. In most cases, it is done by trial and error method. This can also be done by grid search, random search, and Bayesian optimization.

Examples: learning rate, size of neural network, no. of neuron, epoch size, batch size, etc.

CONV Layer

The convolution layer, in short CONV Layer, extracts features from an input image by applying a large number of filters in parallel, which are specific to the dataset and the predictive modeling problem. The input image is the RGB model, which comprises three channels, red, blue, and green. The 3-D volume of neurons performs the convolution operation. The weights for the neuron are nothing but the filters.

Filter

Filter can also be called kernels. Each filter contains a set of weights that are for a specific purpose, like for edge detection, filter for line detection, sharpening, Gaussian blur, etc. The size of the filter is mostly 3×3 or 5×5, but it can also be changed based on the requirement. The filter uses Sliding Window Protocol. Since the size of the filter is smaller than the input image, the filter starts at the start of the input, and then slides across the entire input. The filter acts as a feature detector and sifts the input. It retains the required information in the feature map and removes the irrelevant details. For example, if a line detector is the filter used, upon passing over input, the output contains the information only about the lines that are present in the input.

Feature Map

The result of the convolution layer is a feature map. The feature map is a result of repeated application of the same filter in different patches of the input image. The resulting feature map contains the location and intensity of the detected features of the image. The size of the output feature map is given by

$$\textit{Dimension of the Feature Map} = \frac{W - F + 2P}{S} + 1,$$

where, W – Size of the Input Data (Image)
F – Size of the Filter
P – Padding
S – Stride

Convolution Operation

Convolution operation between the input image and the kernel is the key idea behind the convolution neural network. Convolution operation is a linear operation that involves the dot product of the input image with the kernel. Convolution operation is indicated by "*."

Consider an input image of size 5×5, i.e., $W = 5$, with zero padding, $P = 0$, and the size of the filter is 3×3, $F = 3$ and Stride $S = 1$, based on the formula given above for the dimension of the feature map, upon substituting

$$\text{Dimension of Feature Map} = (5\text{–}3 + 0/1) + 1 = 3.$$

Hence the size of the feature map would be 3×3.

The feature map given above is calculated by sliding the filter across the input and doing dot product.

Similarly, the filter slides next step, and when all the values have been covered horizontally, it then slides downward, i.e., vertically to the beginning of the next row, as shown in Figures 1.11 and 1.12.

This way the convolution process takes place. The output feature map is then passed to the ReLU activation function. This ReLU layer is comprised within the CONV Layer. The purpose of the ReLU function is to remove the linearity in the feature map obtained and make it more meaningful. It removes negative values from the feature map by replacing them with zero.

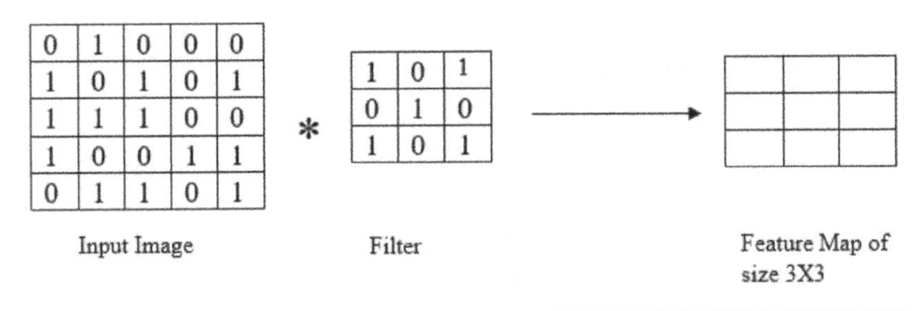

Figure 1.11 Convolution Layer

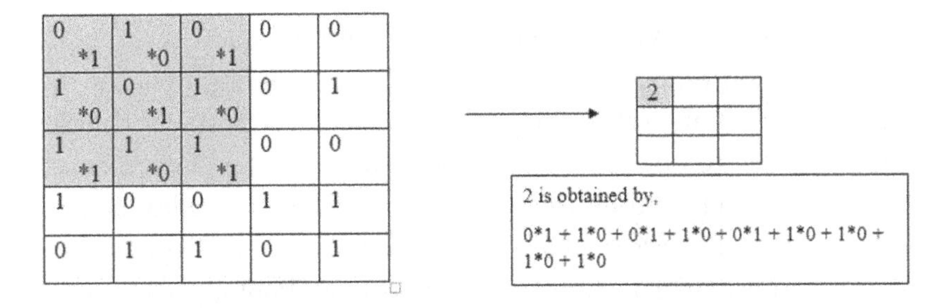

2 is obtained by,
0*1 + 1*0 + 0*1 + 1*0 + 0*1 + 1*0 + 1*0 + 1*0 + 1*0

Since Stride is 1, the filter slides one step.

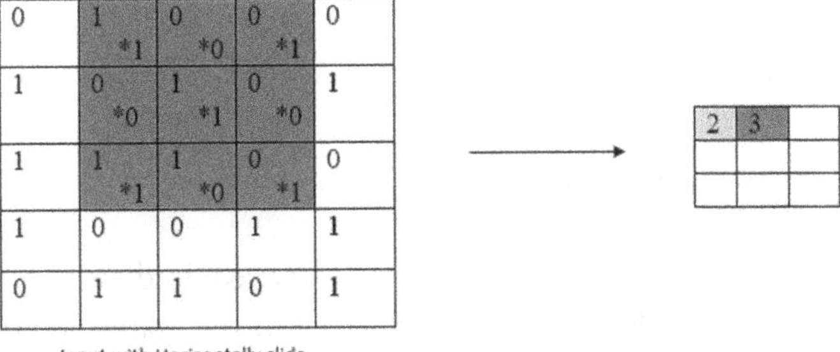

Figure 1.12 Filter Slides throughout Image

Key Points about Convolution Layers and Filters

- The depth of the filter must be same as that of the input to ensure that all the input layers are processed. Depth denotes the number of color channels.
- Deeper layers extract more features but are computationally expensive.
- Filters are automatically initialized on the basis of their normal or Gaussian distribution for each region.
- The number of filters in the initial layers should be less, and the deep hidden layers have twice or the thrice the number of filters in the initial layers.

Pooling Layer

Pooling layer is added either after a single layer of convolutional layer or after two layers of CONV Layer; a single layer of pooling is added. When the images of the dataset are very large, the corresponding feature map also will be of higher dimension. This subsequently increases the computational cost. Pooling layer removes the unwanted parameters, thereby reducing the size of the feature map.

Pooling, also called sub-sampling or down sampling, reduces the size of each output feature map, but pooling layer ensures that the important features are retained. It does down sampling by removing unnecessary details from the feature map. The key idea behind pooling is that the relative location of the feature is enough and the exact location of the feature is less important. Pooling is generally done over 2×2 square values of the feature map.

Key Points about Pooling Layer

- Pooling reduces the spatial size.
- It reduces the parameter size.
- It reduces the computation and the memory required.
- It operates independently on every channel of the feature map.
- Stride decides how the elements are pooled in a way similar to CONV layer.

Types of Pooling

1. **Max Pooling:** Maximum of the four values in the 2×2 square selected from the feature map is the selected value as shown in Figure 1.13.
2. **Average Pooling:** Average of the four values from the 2×2 square values of the feature map is chosen as the final value (see Figure 1.14).
3. **Sum Pooling:** Sum of the four values from the 2×2 square values of the feature map is chosen as the final value (see Figure 1.15).

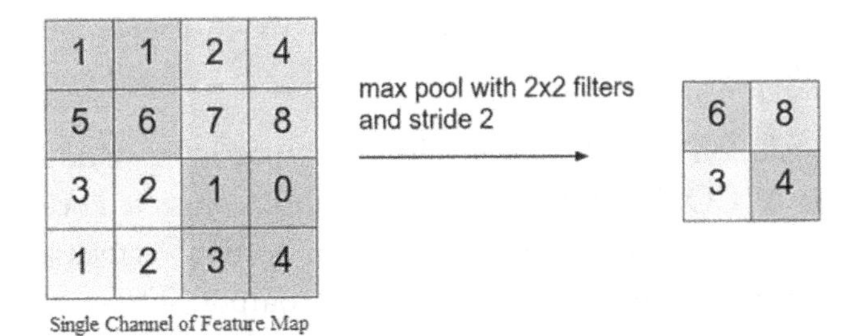

Figure 1.13 Max Pooling

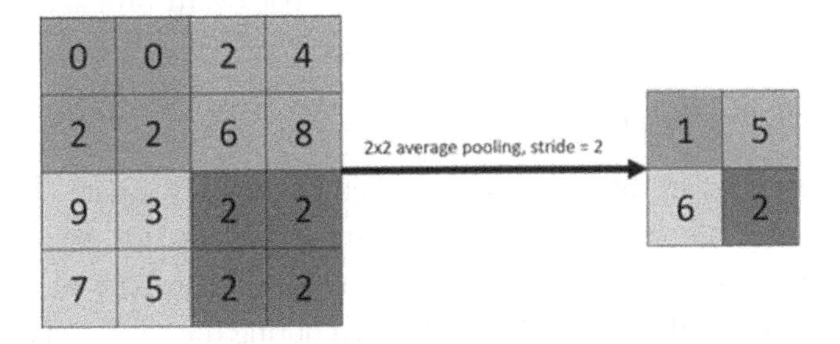

Figure 1.14 Average Pooling

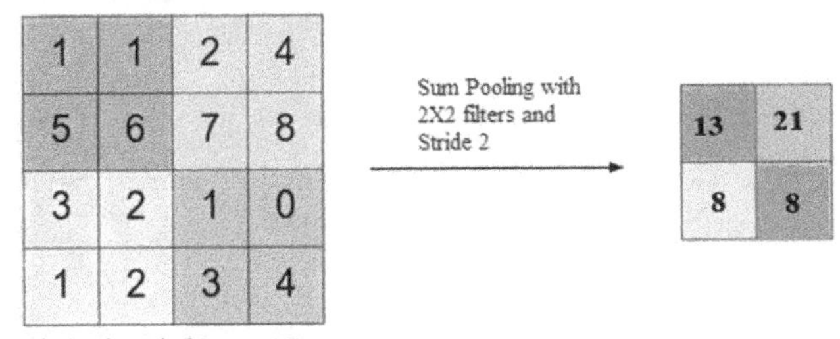

Figure 1.15 Sum Pooling

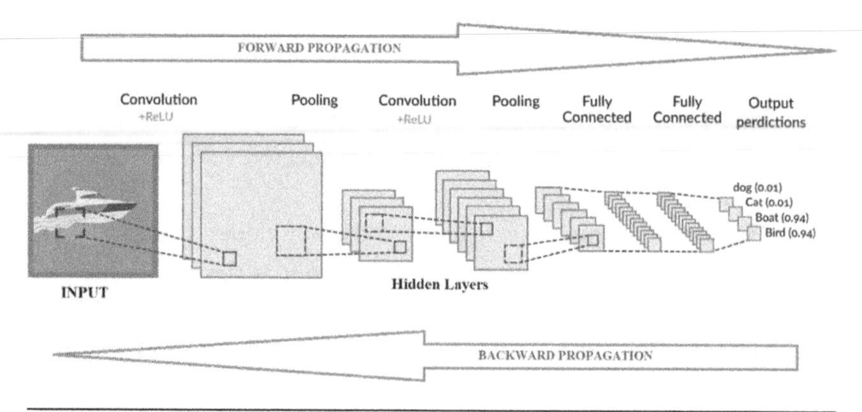

Figure 1.16 Forward Propagation

Forward Propagation

The convolution neural typically has an input layer, hidden layers consisting of CONV and POOL layers, fully connected ANN, and the output layer, as shown in Figure 1.16. Softmax activation function is often preferred for generating the output. The convolution neural network consists of training part and testing part. In the training part, optimization can be done on the basis of the accuracy obtained, which is carried out during the backward propagation.

Calculating the Parameters

Parameters are basically the learnable elements from the data. They are weights that are learnt during the training process, for example see Figure 1.17. These weight matrices are used to contribute to the predictive power of the model that is changed during the back propagation.

	Activation Shape	Activation Size	# Parameters
Input Layer:	(32, 32, 3)	3072	0
CONV1 (f=5, s=1)	(28, 28, 8)	6272	608
POOL1	(14, 14, 8)	1568	0
CONV2 (f=5, s=1)	(10, 10, 16)	1600	3216
POOL2	(5, 5, 16)	400	0
FC3	(120, 1)	120	48120
FC4	(84, 1)	84	10164
Softmax	(10, 1)	10	850

Figure 1.17 Number of Parameters

Steps in calculating the parameters

1. Input layer has nothing to learn. So the parameter count is 0 in input layer.
2. CONV layer: The formula to find the number of parameters learnt from the convolution layer is

$$\textit{No. of Parameters in CONV Layer} = ((m * n * d) + 1) * k),$$

where,
m – width of the filter
n – height of the filter
d – no. of filters in the previous layer
k – no. of filters and 1 indicates the bias term

3. Pooling layer: Learning does not take place in the pooling layer. Hence, the number of parameters is 0.
4. Fully connected layer: The fully connected layer has the highest number of parameters. The formula to calculate the number of parameters in FC layer and the softmax activation function is

$$\textit{No. of Parameters in FC Layer} = (c * p) + 1 * c,$$

where,
c – no. of neurons in the current layer
p – no. of neurons in the previous layer and 1 is the bias term

Activation Shape and Size

Activations maps are just another name for the output feature map of the convolutional layers. The activation shape is nothing but the dimensions of the output of the particular layer in the format (height, width, and no. of channels). Activation size is the product of the height, width, and the number of channels.

The forward propagation computes the result of the learning process and saves the intermediate values that are required for the gradients computation. The gradients computation is used for calculating the loss function and the optimization (see Figure 1.18).

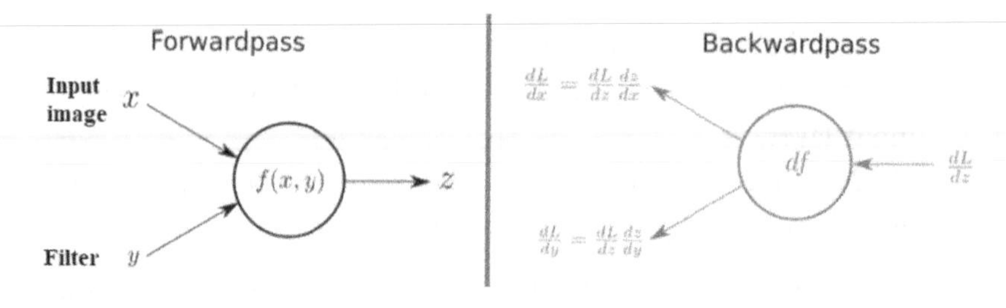

Figure 1.18 Forward and Backward Propagations

Backward Propagation

Once the forward propagation is done and the output is predicted, the loss function or the cost function evaluates the model. If the computed value of the loss function is high, it means that the model has not been trained properly. If the value is low, it denotes that the model is working with good accuracy. The loss function is indicated by L. The loss function is selected on the basis of the application for which the CNN architecture is modeled.

Whenever, L is high, there's a need to back propagate and alter the weights used in each layer in order to optimize the working of the model. The weights here refer to the filter values. Given a CNN and the loss function value, the backward propagation computes the gradients of the loss function with respect to the weights and biases in the network. In gradient computation, differentiation is done, which denotes the rate of change or difference between the expected output and the actual output obtained. It is important to note that the backward propagation of the convolution operation is also a convolution operation with flipped kernels.

In Figure 1.18, $\frac{dz}{dx}$ and $\frac{dz}{dy}$ are local gradients and are calculated and saved in memory during the backward propagation itself. The local gradients are calculated from the intermediate values stored in the forward propagation. By applying the differentiation chain rule and the Jacobean notations, the weights can be altered, and then the model is again forward propagated to find the new output. This process is carried out repeatedly until the required accuracy is obtained. This process can also be called fine-tuning of the CNN.

Optimizers

During the backpropagation, to minimize the value of the loss function, we use optimizers. An optimizer moulds the neural network to attain maximum accuracy. It is guided by the loss function. Adam optimizer is most commonly used for the CNN architecture, since it gives 99.2% accuracy.

Adam optimizer stands for adaptive moment estimation. Adam optimizer adds the gradients of the previous layers to the current layer under the concept of momentum.

Other Ways to Improve the Performance of CNN

While backpropagation and parameters tuning act as the primary method to the performance and the accuracy of the CNN, there are three other proven ways to improve the model, which are described below.

Image Data Augmentation When the amount of data used for training is less, the model may not be able to learn the key features properly. This results in lower accuracy of the model. This can be overcome by the concept of augmentation. Wherein the image data, which are already used for the purpose of training, can be augmented with various operations, like rotation, reflection, tilting, performance of random crop, and alteration of the image properties as shown in Figure 1.19. These images add the count to the dataset, enabling the model to learn better.

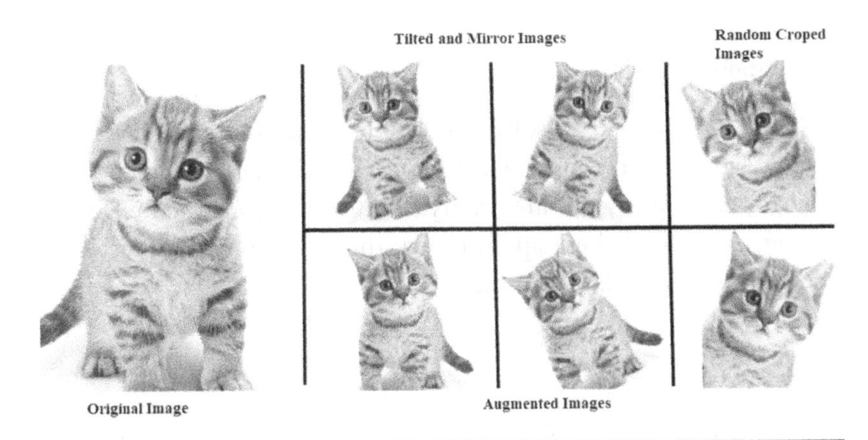

Figure 1.19 Augmentation

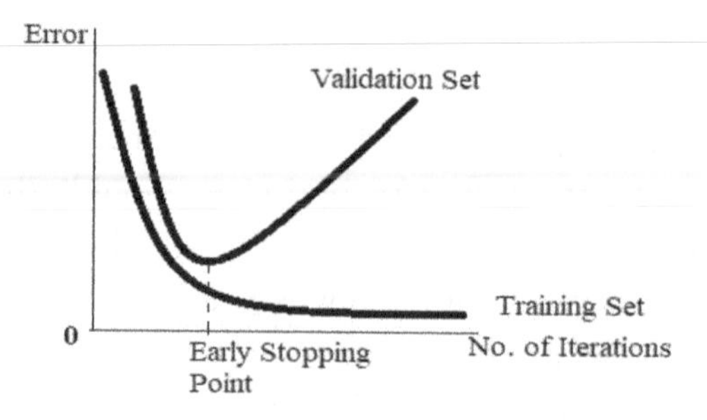

Figure 1.20 Early Stopping

Deeper Hidden Layers Having multiple convolution layers enables better features extraction and better learning. This increases the accuracy of the model. It is also important to note that, better accuracy is provided by a deeper network and not essentially by a wider network. By wider network, we mean large number of filters in a single layer of CNN. The wider layers are better at learning the features of the input images but deeper networks with hierarchical structure are good in generalizing the learning process. This generalization would improve the accuracy for a wide variety of dataset.

Early Stopping As the convolution neural network performs the training process in multiple iterations, it is important to ensure that the model does not start to overfit. To ensure this, the iteration has to be stopped at an early stage when the loss function value is less. To find the ideal early stopping point, the network must be fed with the validation set (see Figure 1.20).

Application of CNN

CNN is mainly used to solve the problems of computer vision like image classification, object localization in an image, segmentation, Object detection in the image, style transfer, reconstruction, colorization, etc. Of all these applications, the most commonly used application is image classification.

Image Classification

Image classification is of two types: binary classification and multi-class classification. Binary classification classifies object between the two given data classes, whereas multiclass classification classifies the object as one among the many data classes.

Binary Classification

Using python, tensorflow and keras libraries, we can classify images using the CNN. Following are the steps to create a classifier:

1. Create a folder for the 2c classes of object in the python.
2. Open the python notebook and import the required libraries, like sequential, Convolution2D, Maxpooling2D, Dense, etc.
3. The next step is to initialize the CNN architecture, using the command *model = Sequential().* Here, model is the name of the architecture and we are creating it in sequential pattern.
4. Add the convolution layer to the convolutional layer, *model.add (Convolution2D(filters=32, input_shape = (64, 64), kernel_size = (3,3,activation = 'relu'))*
   Here, the parameters are no. of filters = 32, size of the filter – 3×3, Size of the input – 64×64 and the activation function is ReLU.
5. Add the pooling Layer.
   model.add(Max_Pooling2D(pool_size = (2, 2)).
   Here, pooling window size is 2×2.
6. Similarly add consecutive layers of convolution layer and pooling layer.
7. After adding the conv and pooling layers, the output is in the form of a feature map. A feature map is a matrix. As we know, the last layer of CNN architecture is a fully connected neural network and it takes only an array of data as input, we convert the matrix into a 1-D structure using flatten.
   model.add(Flatten())
8. Add fully connected neural network to the sequential network.
   model.add(Dense(units = 128, activation = 'relu'), Here, units indicate the number of neurons.
   Output layer:
   model.add(Dense(units = 1, activation = 'sigmoid')
   Here, the activation used is Sigmoid, because the application is binary classification and the output should either be 1 (if class 1 object is present in the image) or 0 (if class 2 object is present).

9. Now that the model has been developed, it can be compiled. At the time of compilation, the loss function to be used and the optimizer has to be specified.
   model.compile(optimize = 'adam', metrics = ['accuracy'], loss = 'binary_crossentropy')
   Metrics is a list of elements, and multiple metrics can also be optimized.
10. Now, the model is developed. Training Data has to be fed to the network.
    The data has to be rescaled to a fixed size and pre-processed using the keras library.
    from keras.preprocessing.image import ImageDataGenerator
    train_datagen = ImageDataGenerator(rescale = 1./255, zoom_range = 0.2)
11. Feeding the training data and test data:
    training_set=train_datagen.flow_from_directory('data/training_set', target_size=(64,64),batch_size=32, class_mode = 'binary')
    test_set = test_datagen.flow_from_directory('data/test_set', target_size = (64,64), batch_size = 32,class_mode = 'binary')
    Here, directory refers to location where the dataset is stored, and target size indicates the size to which the data would be resized. Batch size indicates the number of data to be passed for each iteration and class mode here is binary, since the application is binary classification.
12. Training the model is done by,
    model.fit_generator(training_set, samples_per_epoch = 2000, nb_epoch = 15, validation_data = test_set, nb_val_samples = 200)
    Here, generator generates the training process with training set, samples per epoch indicates the number of batches i which the data have to be fed to the network, nb_epoch indicates the total number of iteration, validation data indicates the set of data, which is used for testing purpose, and nb_val_sample indicates the number of samples to yield before stopping each iteration.
13. To view the summary of the model,
    model.summary()
    This command gives the accuracy obtained and the number of parameters generated by each layer.

RCNN and Object Detection

Object classification algorithm identifies whether a particular class of object is present in a given input image or not. If the object is present, it is equally important to identify the location of the object in the image. Image detection is done to identify whether an object is present and then locate the object(s) with the help of a bounding box. Object detection is important for various applications, such as vehicle detection in self-driving cars, human activity recognition in the surveillance feed, etc. But to implement this, the normal CNN architecture is not efficient enough to compute the bounding boxes. The example of bounding box is shown in Figure 1.21.

The main reason for the regular CNN architecture to be not able to compute the bounding box is that the sliding window pattern of filtering is not efficient enough to identify the bounding box. One more difficulty with object detection is that there could be any number of objects of interest, and each might be of different aspect ratios and different locations. In order to identify the location and identify the number of objects, using the traditional CNN architecture would be computationally very expensive and less efficient. Hence, instead of using the sliding window protocol, selective search protocol is used. The various CNN-based architectures to do object detection are R-CNN, Fast R-CNN, and Faster R-CNN and YOLO.

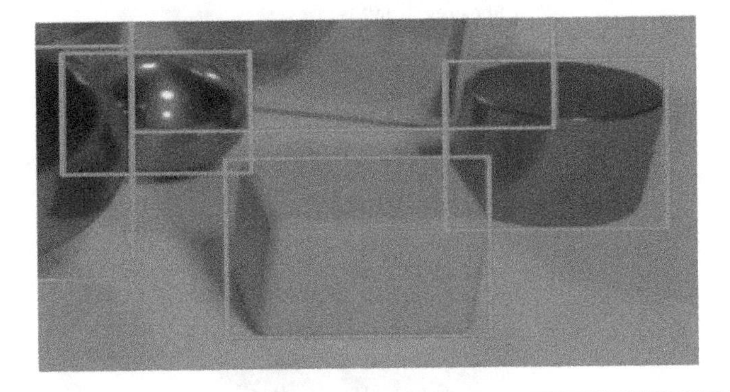

Figure 1.21 Bounding Boxes

Region-Based Convolutional Neural Network (R-CNN)

Instead of selecting large number of sub-regions from image, R-CNN uses the selective search, to extract just to 2000 sub-regions.

The Selective Search Protocol is given below:

a. Generate candidate sub-segmented candidate region.
b. Using Greedy algorithm, the similar regions of the image are recursively combined to a larger image.
c. Use the combined region to produce the complete image.

The image proposals are warped into 4096 dimension squares and are fed to the CNN. The CONV Net performs the feature extraction process and then passes the identified features to the support vector machine for classification and the offset (Location information) to the bounding box regressor.

Limitations of R-CNN:

- Since selective search algorithm does not learn, there could be possibilities of poor candidate region proposals.
- It takes around 47 seconds for the R-CNN to process a test image.
- Training is also slow, since it takes 2000 candidates from each image.
- It is computationally expensive.

The process of the R-CNN is shown in Figure 1.22

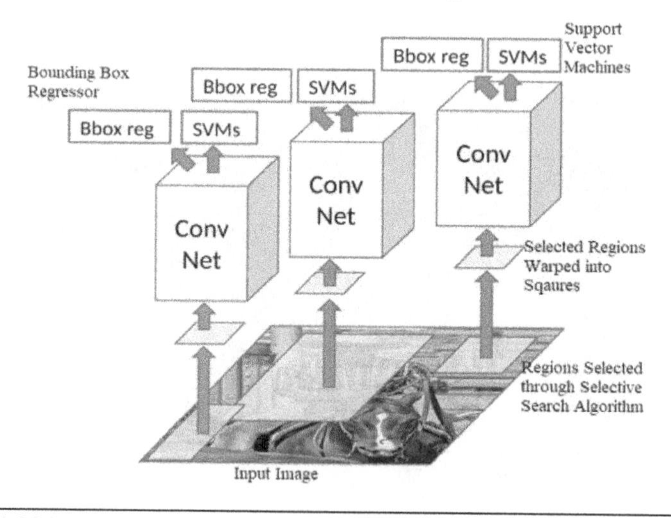

Figure 1.22 R-CNN

Fast R-CNN

To overcome the limitations of the R-CNN, the Fast R-CNN was introduced. It is faster than the R-CNN, and hence the name. Fast R-CNN does not use selective search algorithm. In the fast R-CNN architecture, the input image is given as input to the CNN architecture. The CNN layer extracts the feature and produces the feature map. It is important to note that the object proposals are extracted using the selective search algorithm from the feature map. Using the object proposal, the co-ordinates of the object region are identified in the feature map and stored as region of interests projection. The pooling layers, with the help of region of interests projections, pool the feature map to fixed length feature vectors of the objects of interest as object proposals, and feeds them to the fully connected layers and the softmax for the classification tasks. The offsets of the identified regions are given to bounding box regressor to create the bounding box around the object.

The architecture of Fast R-CNN is shown in Figure 1.23.

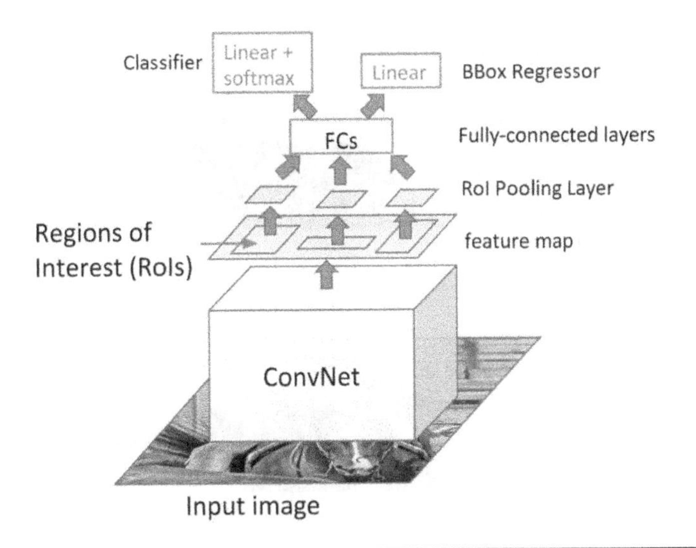

Figure 1.23 Fast R-CNN

Faster R-CNN

Though Fast R-CNN is faster than the R-CNN, it still uses the selective search algorithm, which is a time consuming process. Similar to the Fast R-CNN, in Faster R-CNN also, the input image is directly fed into the CNN. Then the output feature map is separately fed into a pooling layer and a special Region Proposal Neural Network. This network generates the object proposals and is then passed to the region of interest pooling layer. Now that the pooling layer processes on both the feature map and the object proposals, it identifies the regions of interest, warps them to squares, and passes these to the classifier. The classifier then classifies the object and show cases the location using the bounding box. The use of a separate region proposal network instead of selective search algorithm makes fast R-CNN even faster than the other two variants (see Figures 1.24 and 1.25).

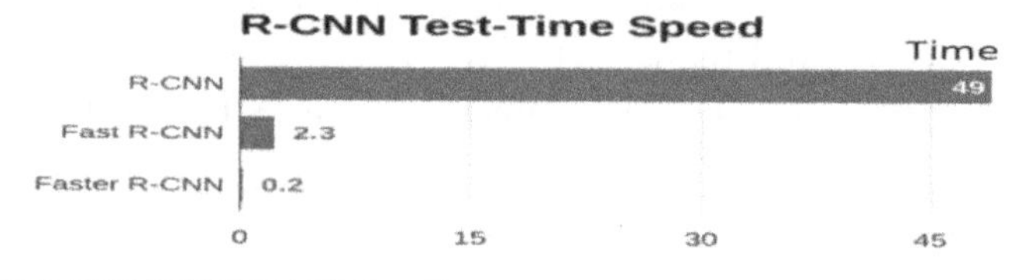

Figure 1.24 Comparison of Execution Speed

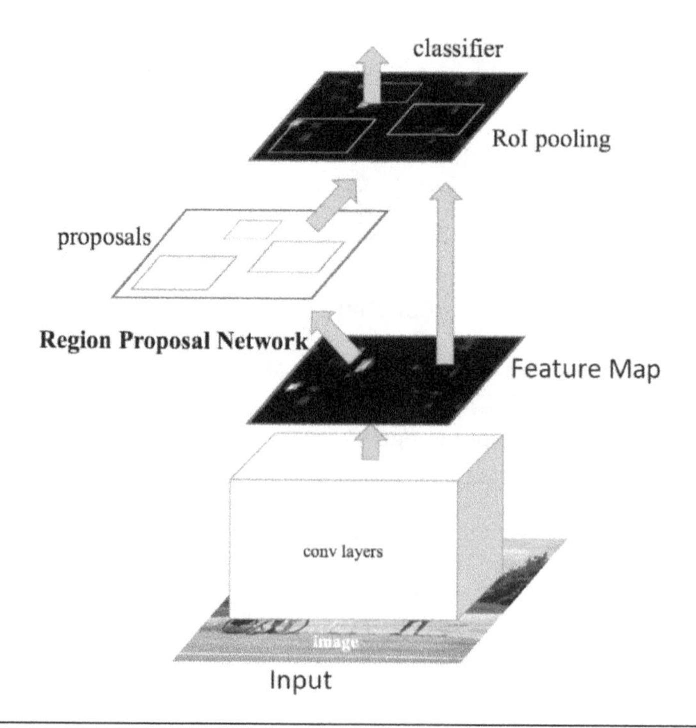

Figure 1.25 Faster R-CNN

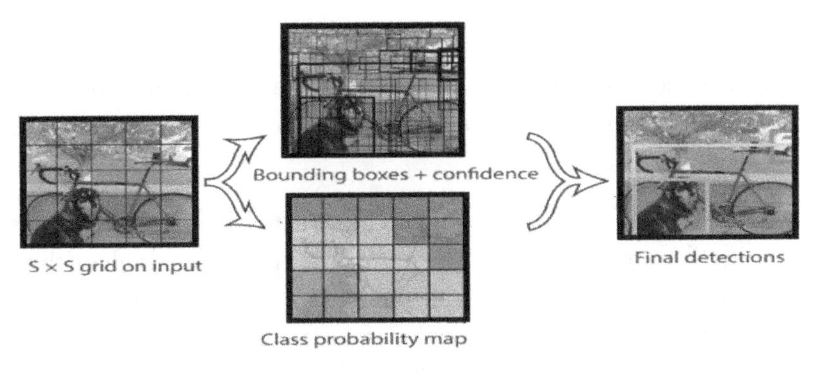

Figure 1.26 YOLO

You Only Look Once (YOLO)

The R-CNN architecture variants use separate algorithms and networks to capture the location of the object. YOLO does not use any additional network for boundary box regression. Instead, a single convolution network is used to perform the bounding boxes and the class probabilities.

When an input image is fed into the YOLO architecture, it splits the image into grids of $S{\times}S$ dimension. S is selected based on the size of image. Within each grid, we take n boundary boxes. The boundary boxes predict the class probability and the offset values. The predicted probability is compared with the threshold and if it is above the threshold value, it is located as a particular object (see Figure 1.26).

Transfer Learning

Transfer Learning is a machine and DL technique that uses knowledge gained in the previously trained model like features, weights, etc. for training new model. Transfer learning aids to better accuracy and overcome the difficulties in training such as setting optimal hyper parameter values, overcoming training data shortage, choosing the ideal optimization technique, etc.

Transfer Learning for CNN

Most CNN Architectures are not developed from scratch, instead the weights and the layers of the previously modeled architectures that are well trained over a larger dataset and have attained higher efficiency are used in modeling CONV net for a new application.

Transfer learning can be applied in the following two ways for the CNN architecture:

1. **Pre-trained Model as Feature Extractor:** This method uses the feature extraction layers of the pre-trained algorithm, but the classification or the prediction process is done by the layers that are newly modeled for the task. This can also be called as freezing the convolutional layer.
2. **Fine-Tuning Pre-trained Models:** This method fine tunes the feature extraction layers of the existing pre-trained models and may or may not use the final fully connected layers of the pre-trained model.

Fine-Tuning or Freezing?

It is important to decide the type of the transfer learning to be used while developing the CNN architecture. Few points to identify the ideal method are:

- When the dataset for the new task is less, it is better not to fine-tune.
- When the dataset for the new task is large, but similar to the pre-trained model's dataset, fine tuning might only produce a small change in accuracy.
- When the dataset for the new task is less, but is different from pre-trained dataset, it is better to train only the classifier or final output layer.
- When the dataset for the new task is large, and different from the pre-trained model's dataset, fine-tuning but feature extraction and the classifier or developing a new CNN architecture could yield better result.

Neural Style Transfer (NST)

Neural style transfer is one of the applications of image styling and transformation technique. NST can be performed by the CNN.

$$\textit{Input Image} + \textit{Style Image} \rightarrow \textit{Styled output image}$$

Steps in NST:

1. Both the input x and the style image a are converted to equal dimensions.
2. A pre-trained convolution neural network can be used for the style transfer. Transfer learning is done here. Ex. VGG16 architecture can be used.
3. The style components like color, basic shapes, and patterns are extracted from the style image, denoted as $S(a)$
4. The content component featuring the actual object and features in the input image, other than color and styling elements, are extracted, denoted as $C(x)$.
5. After extracting the required components, they are fed into the CNN and an output styled image y is obtained.
6. The losses for NST are specific. They are
    1. Content Loss: The difference in content between the input and the styled image, denoted as $|C(x) - C(y)|$.
    2. Style Loss: The difference in style factors between the style image and styled image, denoted as $|S(a) - S(y)|$.
    3. Total Loss: The total loss is denoted by L.

    $$L(y) = |C(x) - C(y)| + |S(a) - S(y)|$$

    The total loss is a regularization loss.
7. Once the loss function is found, optimization has to be done through back propagation, until the loss is minimized. For NST, with large number of parameters, Limited-Memory Broyden–Fletcher–Goldfarb–Shanno (L-BFGS) optimization algorithm works the best.

The NST is done by various mobile applications for photography editing and art works. One such application is Prisma. Figure 1.27 shows an example of NST.

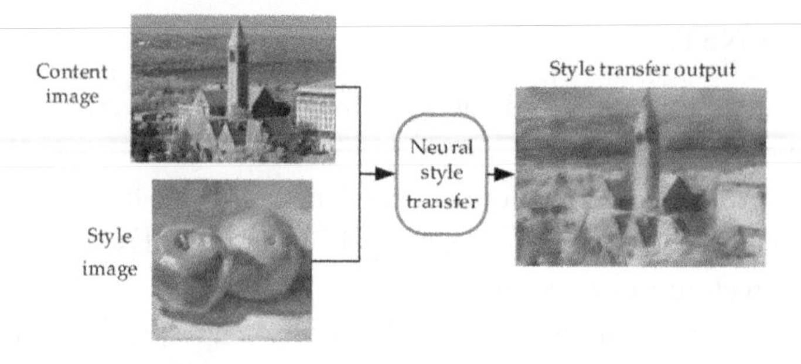

Figure 1.27 Neural Style Transfer

2

Trends and Transition in the Machine Learning (ML) Space

SRUTHI ANAND[1], N. SUSILA[1], AND USHA SAKTHIVEL[2]

[1]Department of Information Technology, Sri Krishna College of Engineering and Technology, Coimbatore, India

[2]Dean, Dept. of Research and Innovation, Rajeswari College of Engineering, Bengaluru, India & Prof. & Head, Dept. of Computer Science & Engineering, Visvesvaraya Technological University, Belagavi, Karnataka, India

Contents

DOI: 10.1201/9781003119838-2

Introduction to Machine Learning

Machine learning (ML) is a subset of artificial intelligence. ML is a branch of study where algorithms are not explicitly needed to be programmed, rather it directly allows computers to train on a particular set of data and the use of statistical analysis to predict the output. Using these, models are built which in turn enable computer systems to involve in the process of decision-making. The concept of ML is based on the idea that the computer programs can access and get trained by themselves which they use to learn for future processes. ML is one of the most commonly used technology floating around that can be used in any domain. The technology as perceived by Alan

Turing that is "Can machines think" is now transformed to "Can machines do what humans think." [1] The ability to solve complex problems is not only the task of humans, but can also be embedded in the systems giving them the capability to think and formulate solutions based on analytical models.

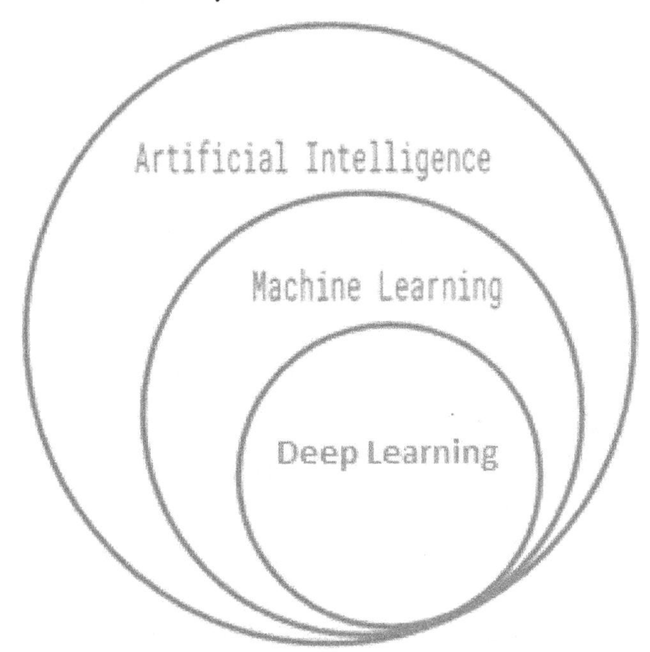

The field of ML has benefited a number of applications, like optical character recognition, that help in conversion of texts into movable types, user behavior prediction, facial recognition, and recommendation systems for identification of the interests of the users in different social media, as shown in Figure 2.1.

As quoted by Tom Mitchell – "A computer program is said to learn from experience E with respect to some class of tasks T and performance measure P, if its performance at tasks in T, as measured by P improves with experience E." For instance, let us assume that we have a group of animals and we have to classify them and give the count. So how do we proceed? The answer is – for a human eye, it is easy to identify the animals based on different features, like their eyes, shape, color, etc. But for a machine how can it understand? The machine must be trained to identify the different features so that it can classify them. ML now deals with processing different kinds of data and identifying the features based on which the prediction can

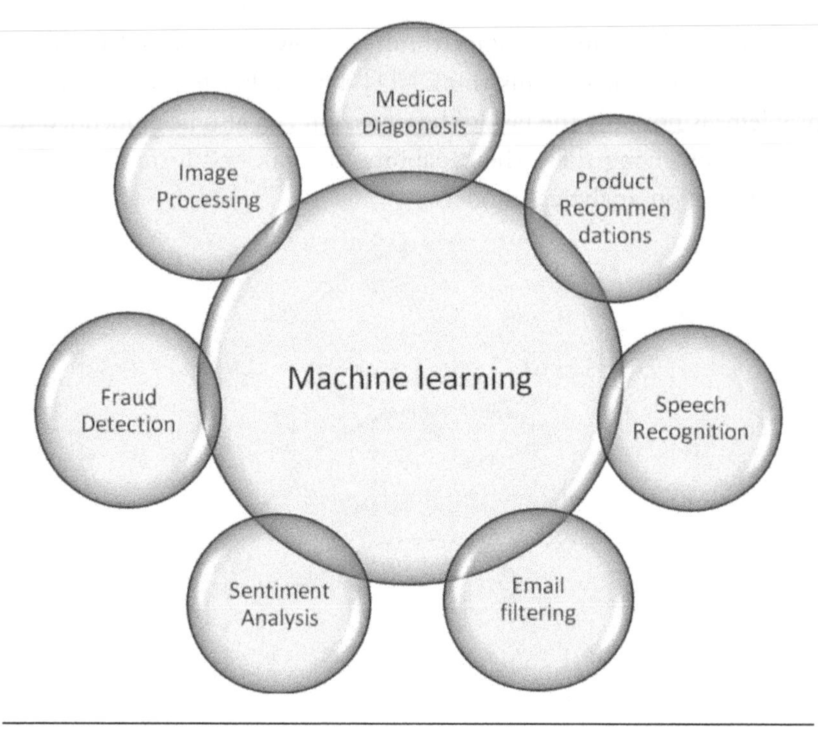

Figure 2.1 Applications of ML

be made. ML helps in many complex tasks, like object recognition, summarization of the predicted object, classification, and clustering, and recommends and suggests the kind of systems to be used for different purposes. ML techniques may be used to gain useful insights from the data that can be used for decision-making.

Motivations for Machine Learning Algorithms

Have we ever thought of how these algorithms have evolved. Let's just take a look at things that are influencing us in our daily lives. Considering the way we handle Google, the search engine processes more than 40,000 searches per second. Have we thought how Google brings us the most accurate result? There are many other examples that we can quote, such as the prediction of a route in our Google Maps, the Spam folder, the options of autocomplete, etc. Can we imagine what if we do not have a system like this? The answer for all these is machine learning. The key idea behind this is given the data, we can extract useful information from it. The idea behind ML is to reduce the tasks of humans in terms of decision-making in complex task processing, build ecofriendly systems by receiving feedbacks instantly and develop technology that can work like humans.

The great motivation factor toward ML algorithms is the thought to develop some technologies that could be derived from human behavior. These technologies have been increasingly used in every industry without realizing that they are being used. ML has evolved from pattern recognition where computers were able to learn with the help of patterns without being explicitly programmed. With the rise of ML algorithms, complex calculations can be applied to big data for faster processing.

Data Analytics and Machine Learning

Data analytics is the field of science that is used in analyzing raw data and draw conclusions from it. It draws patterns and is used in many specialized systems that are integrated with ML algorithms. Data analytics can help to quantify the data and track them to make smarter decisions. Data analytics works with answering the queries based on some existing data, while ML is not focused toward the queries, rather it works with datasets in unstructured ways to get the output. ML is much more toward asking the right questions, and data analytics provides a way to answer those questions and find actionable data. Thus both data analytics and ML are interconnected, with data analytics having micro scope in specific applications and ML having macro scope.

Let's define some important terms:

Artificial Intelligence – The ability of the machines to mimic human behavior and exhibit the traits that are associated with the human mind, such as self-learning and problem solving [2].

Machine Learning – A subset of artificial intelligence that involves the study of different algorithms that learns and improves automatically with experience. ML algorithms are broadly classified into four types (Figure 2.2):

1. Supervised Learning
2. Unsupervised Learning

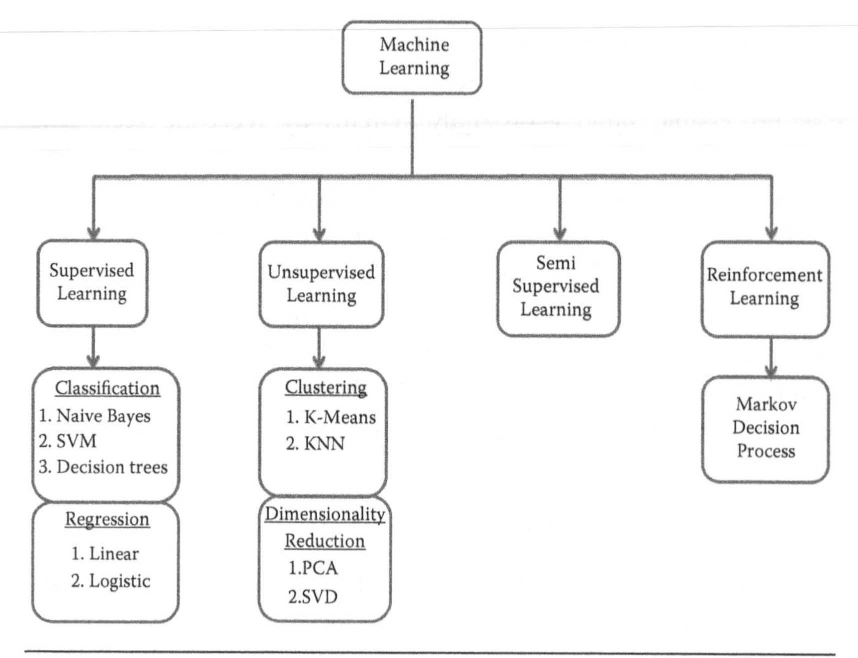

Figure 2.2 Categories of ML

3. Semi-Supervised Learning
4. Reinforcement Learning

Supervised Learning

A model is built using a set of input data that are trained called training data to make the system predict the output. The predicted data may not always be correct and in this case, the data are still trained until the model reaches a desired level of accuracy. The term supervised learning is analogous to a teacher supervising the activities of the students [3]. This learning method takes an input data (*X*) and is applied to an algorithm to predict the output (*Y*). In short, it is denoted as

$$Y = \text{Function}(X)$$

Supervised learning techniques are further classified into two, namely:

Classification: The process of classification is to predict if the output belongs to a particular category viz {true or false}, {red, blue or green}, {positive or negative} etc.

Examples include classifying the emails into spam and not spam, classifying documents as confidential or not, predicting the user behavior and classifying as churn or not, classifying a tumor if it is malignant or benign.

Regression: The process of regression is identifying and predicting continuous values based on either a single value or a set of values.

For example, regression can be used to predict the salary of an employee based on age, calculate the BMI based on the weight, and calculate the price of a car based on its year and the number of kilometers it has run, etc.

Classification

Classification is a method to understand, recognize, and group based on a particular label or category. There are a number of algorithms for classification, such as naive Bayes classification, support vector machine, decision trees, and random forests. These algorithms take some input data and are trained such that the subsequent data may fit into one of the categories [4]. Let us analyze and understand these algorithms in the following descriptions:

Naive Bayes (NB)

It is a classification algorithm that is based on the Bayes theorem, by the use of conditional property. Bayes theorem helps to determine the probability of an event by considering the probability of another event that has already occurred. Mathematically, it is written as:

$$P(A|B) = P(B|A).\ P(A)/P(B)$$

The probability of A can be determined assuming B has occurred. Here B is the evidence and A is the hypothesis. These classification algorithms can be used for categorizing a class of data, given the features as inputs. This algorithm works by taking into consideration that all the features and the input variables are not dependent on each other. This algorithm treats all the features independently to obtain the result. NB can be used in text analysis and pattern recognitions [5].

Few examples that can be predicted using NB are as follows:

1. Finding out if a person has diabetes or not considering the age, diet, pressure, gender, etc.
2. Identifying the weather conditions based on the temperatures, humidity, etc.
3. Identifying a particular climatic condition if it is suitable for a particular sport.

There are different classifications of NB such as Gaussian NB, Multidimensional NB, and Bernoulli NB used in various dimensions. This algorithm is generally very fast and can be used for both binary class and multidimensional classifications. This NB classification algorithm can be used to train small datasets and make predictions on large volumes of data. Since the algorithm takes in all features independently to analyze the data, this process could be more complex if it has multiple attributes [6].

Support Vector Machine (SVM) Classifier

SVM are algorithms that can be used to solve classification and regression problems, but it is used for classification in majority of the cases.

Types of Learners

It is a simple algorithm, where the generalization of algorithm takes place after a query is raised. Though they consume less time in training, they result in increase of prediction time.

Example: K-Nearest neighbor

Eager Learners

In this algorithm, the execution happens immediately and produces desirable output. Though they consume a lot of time in training, they result in less predicting time. They can be optimized.

Example: NB, Decision trees
Example for Classification:

Document Classification

Document classification is a process of defining documents and separating them into distinct classes based on their contents. This classification will be helpful for people who deal with enormous amount of contents.

The words from a document can be used as "features" to help in the prediction of the classification of a document.

Illustration

Let's consider two reference classes.

Reference class 1	Reference class 2
Welcome to machine learning	Happy coding

To classify these documents, we start by taking all of the words in the two reference documents and creating a table or vector from these words.

(Welcome,to,machine,learning,happy,coding) class

We can create a vector by assigning a number 1 if the word exists in the training document and number 0 if it doesn't exist, for each of the training documents.

Welcome	To	machine	learning	Happy	Coding	
1	1	1	1	0	0	Class 1
0	0	0	0	1	1	Class 2

Now let's assume a new document comes with the tag,

"Welcome to python learning," for this document, the word vector becomes like

Welcome	to	machine	learning	Happy	Coding	
1	1	0	1	0	0	Unknown class

If we compare this vector for the document of unknown class to the vectors representing our two reference document classes, it closely resembles the vector for class 1 document,

Unknown class: <1,1,0,1,0,0>

CLASS 1: <1,1,1,1,0,0> (matching vectors: 5)
CLASS 2: <0,0,0,0,1,1> (matching vectors: 1)

We can possibly label the data as class 1 (maximum number of matching vectors). This is one of the examples of statistical natural language processing method.

Logistic Regression

This is a technique of ML algorithm, where one or more independent variables results in a desired output. In this case, possibly we have only two outcomes. The ultimate aim of regression is to achieve best fit. It is based on predictive analysis. In this, we can assume the probability with the help of log function.

Practical Use Cases

Weather Prediction

Weather prediction can be done using logistic regression. The process begins with analyzing the previous data and records of weather report, thereby predicting the desired weather specification for the given day. But with the help of logistic regression, we can only predict whether it will happen or not [7].

Patient's Health Monitoring

If a patient is admitted in hospital, his health is to be monitored. The severity of illness can be predicted with the help of his history in medical report. Thereby, we can predict if his severity is high/low of any disease. We can also predict risk factors of a specific disease he is suffering from.

Unsupervised Learning

Unsupervised learning is learning algorithms that do not require labeled responses and can draw their own references. Cluster analysis is the commonly used unsupervised learning algorithm that takes Euclidean or probabilistic distance as a metric. Clustering algorithms include the following:

This clustering creates a multilevel cluster hierarchy by building cluster trees.

It is the simplest of unsupervised algorithm. As the name suggests, clustering divides the data based on certain patterns. These unsupervised algorithms derive conclusions from the input vectors from datasets without using labeled outcome. The process is partitioning the data into distinct clusters based on the means that is the centroid of a cluster. Unlike supervised learning, the outcome is not labeled; instead, it generally groups the data. They are represented by data points, and the cluster that is formed must be relevant.

Gaussian Mixture Models

This includes modeling the clusters using an approach known as soft clustering. They are also probabilistic models that distribute the data points using the soft computing approach. Gaussian mixture models are like Gaussian distribution or the normal distribution. Gaussian considers both mean and variance to update centroid, whereas k-means consider only the mean.

Hidden Markov Models

This model uses observed data to recover the sequence of states. These are used in bioinformatics, data mining, and pattern mining.

Unsupervised learning algorithms allow users to perform more complex tasks compared to supervised learning.

Unsupervised learning is unpredictable compared with other learning methods. Unsupervised learning algorithms include clustering, anomaly detection, neural networks, etc.

Types of Unsupervised Learning

Unsupervised learning problems can be divided into clustering and association.

1. **Clustering**

Clustering technique deals with observing certain structure or pattern from uncategorized data. They process the data by finding the clusters from the dataset. They also consider the granularity and make adjustments based on it.

There are different types of clustering:

1.1 Exclusive (partitioning)

In this method, grouping of data is done belonging to one cluster only.

Example: k-means

1.2 Agglomerative

Here, all the data are clusters. In order to reduce the number of clusters that are formed, union operation is performed.

Example: Hierarchical clustering

1.3 Overlapping

Here fuzzy datasets help in clustering the data. There may be two or more clusters associated with separate degrees of membership. Here, the data are associated with an appropriate membership value.

Example: Fuzzy C-Means

1.4 Probabilistic

This technique uses probability distribution to create the clusters.

Example: Following keywords

"man's shoe."
"women's shoe."
"women's glove."
"man's glove."

These can be clustered into two categories "shoe" and "glove" or "man" and "women."

Clusters can also be typed into:

Hierarchical clustering
k-means clustering
K-NN (k nearest neighbors)
Principal Component Analysis
Singular Value Decomposition
Independent Component Analysis

2. **Association**

This enables association among the data that are available inside the datasets. This technique is used to discover relationships between the variables in very large databases.

Examples:

The cancer patients are grouped by their gene expressions.
Groups of shopper based on their browsing and purchasing histories.
Movies can be grouped by the rating given by the movie viewers.

Applications of Unsupervised Learning

1. **Anomaly Detection**
   Anomaly detection is used to identify the uncommon items, events, or observations from the normal data. In this case, the system trains with large data which are normal instances. So, in an unusual instance, the machine detects whether there is some unusual behavior and claims it to be anomaly.
   One example is credit card fraud detection, which can be solved using anomaly detection techniques in ML. Here, the system detects any unknown credit card transactions, which help in preventing fraudulent activities.
2. **Clustering**
   Clustering is the process of grouping the input dataset into different clusters or groups based on the criteria. Clustering can be done when the exact information about the clusters is not known. For example, in a supermarket, we can see how different items are grouped and arranged. Also, e-shopping websites like Amazon and Flipkart use clustering algorithms to group their products according to the users' needs.
3. **Visualization**
   Visualization is the process of creating diagrams, images, graphs, charts, etc., to deliver and communicate a piece of information.
   For example, a football coach has some data of team's performance in a tournament, to find all the statistics about the matches quickly. We can feed complex and unlabeled data to some visualization algorithm.

These algorithms can be represented as a 2-D or 3-D representation of the data as output in a plotted graph, which helps in determining the winning formula (chances), correcting previous mistakes, and winning the trophy.

4. **Dimensionality Reduction**

   Dimensionality reduction is the process of reducing the number of random variables by using a set of principal variables. The objective of this technique is to simplify the data without losing too much information.

   One of the methods to perform dimensionality reduction is feature extraction, where all correlated features are merged to one. This is a good approach as it can reduce the redundant data and then can be trained for better accuracy of the system.

   Reducing the dimensionality may result in loss of some information. Even if it speeds up the training, it will make the system perform slightly worse.

   So, one should use dimensionality reduction only if the training is very slow. Otherwise, use the original data.

5. **Social Media**

   In social media, we use unsupervised learning techniques to detect frauds and to learn and detect the categories of people based on different criterias.

   Consider Facebook, for example, which uses face recognition; it helps in identifying the people around the globe by identifying each and every minute details of people living in different geographical areas. Also, the news feed section is filtered on the basis of their interests and likings.

Process of Machine Learning

There are seven different steps involved in the ML process (See Figure 2.3):

1. Gathering Input Data
2. Data Preparation
3. Choosing a `Model
4. Training

Figure 2.3 The seven steps of the ML process

5. Evaluation
6. Parameter Tuning
7. Make Predictions

Gathering Input Data

This is the first step of ML process. In this step, we collect the data required for processing. This includes both the quality and the quantity, which in turn are dependent on how good the model can be. For example, if we want to identify a particular object, the first step would be to identify and gather inputs based on the color, shape, size, texture, etc. Incorrect data would lead to incorrect representations and direct the model in a wrong way.

Data Preparation

After gathering the data, the next step is to prepare our data in terms of ordering the data and removing the skews. This means our data must be prepared in such a way that it is ready to be trained for the model. The next important part of data preparation is to break the data into training and testing data, where training data would contribute to 80% and testing data would be 20%, which are used for the evaluation. This step also concentrates in removing duplicates, discarding incorrect data, and fine-tuning for better process.

Choosing a Model

Next is to identify the model that chooses the problem and the kind of learning techniques to be employed. We already know that we have a number of models that can be used for different purposes and it is necessary to choose the right one.

Training

As the name suggests, training is the phase where our model originally starts learning. The 80% of the data that we have allocated for training are used to teach our model. To train a model, more experimentation must be done in order to implement. The usage of weights, biases, and mathematical models help in making predictions. Training is highly successful if the output is satisfying and the outputs are obtained for different sets of input.

Evaluation

This phase actually identifies if the model is able to achieve proficiency and accuracy. Here the model may come across various situations, which it would not have witnessed in its training phase. Evaluation step helps to check if the goals are met, especially using the data that were not used for training. This represents the performance of the model in the real world.

Parameter Tuning

In this phase, we can tune the parameters by making certain modifications and check if we can improve the training to get better results. This may be the inclusion of some other parameters that were not considered in the training and measure the performance. Another way is to run our dataset for multiple times and gain more accuracy.

Prediction

The final step of ML process is prediction. This phase makes the model ready for practical applications. This is the stage where the model starts to perform its own conclusions without human interventions. This step is actually the result of what a human is expected to do. The major difference between human predictions and model prediction is that the models can process huge amounts of data simultaneously, unlike humans. This process helps in reducing the human's burden for decision-making process and arriving at conclusions [6].

Machine Learning – Practical Use Cases

We have discussed the basics of ML above. Now there are a number of trends that are making way in today's market with the technology. As reported by Statista, $28.5 billion were allocated in the technology of

ML during the first quarter of 2019, and we cannot even imagine the increase its use currently [7]. A great number of ML technologies will have a greater impact altering the basis of industries around the globe. These include Regulation of Digital Data, Voice Assistants, Marketing, Cyber Security, and just simply to every kind of industries [8]. As noted, there are a wide variety of applications of ML in the fields like banking, agriculture, healthcare, automobiles, and finance systems.

ML in Banking

The usage of ML banking applications has revolutionized the financial world, making it better for people to use. The major areas are to make credit decisions, fraud prevention, and personal assistants. The banks usually collect a huge amount of data, by which they can derive greater rules that help for credit scoring [7]. The data available are trained using ML algorithms and eventually improve to make decisions on applicants who are subjected to more credit scores. The increase in automation has paved a way toward these technologies where the risk involved can also be assessed in a few minutes, much improved than human efforts. Every financial sector is very much prone to frauds and preventing the frauds is a great deal in today's world. To detect frauds, the patterns of spending and the location of the cardholder or the customer are analyzed and anomalies are detected. In addition, the system can also find any kind of suspicious activities that happen by considering the deviations or abnormalities in the patterns [9].

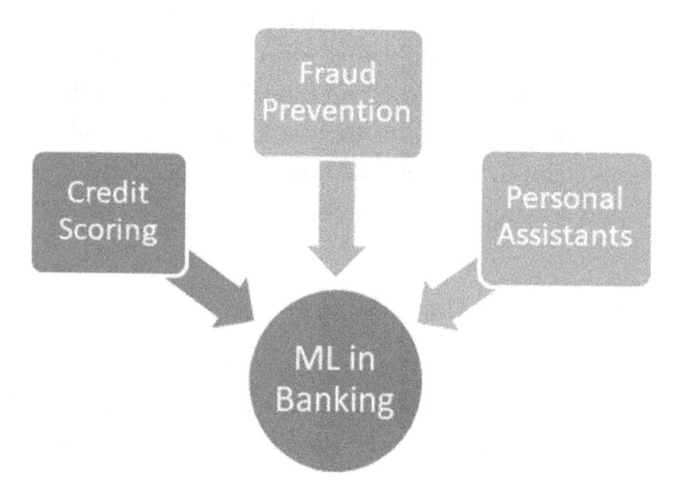

ML in Agriculture

The role of ML in Agriculture plays a pivotal role in our society as Agriculture is one of the predominant fields in the global economy. The lifestyle that we live need a great improvement and it is all determined by the type of food we eat. Promoting and preserving our economy is very much important. Thus growing healthy seeds and preserving the crops, determining what kind of crop to be harvested can all be analyzed by the use of ML algorithms. Many retailers use this technology to create better crops. The other methodologies which can be employed in the field of agriculture are determining the nature of the crops to get the best yield, find bugs and diseases associated with the crops and also to predict when to water the crops. It also helps to predict bug hunters. Agriculture robots are designed to manage and carry out all essential activities from seed sowing, watering, monitor crops and detect diseases. The major part in employing AI and ML is to identify the soil, weather conditions and also nurture the crops that will give good production and make the crops healthy [10].

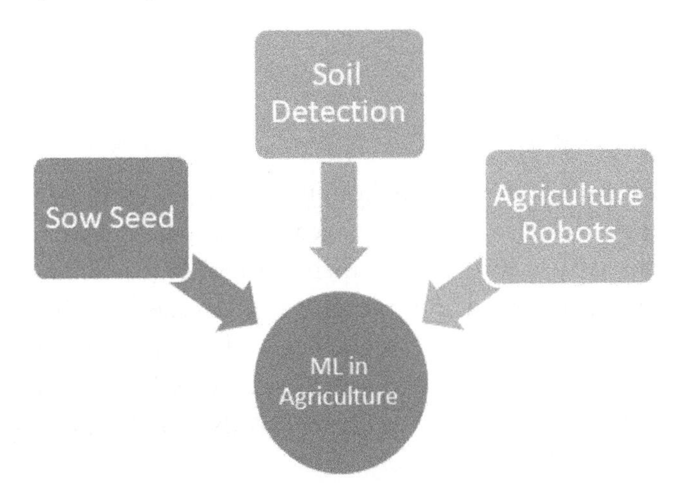

ML in Health Care

The application of ML in the field of healthcare helps patients to analyze and predict the data in prior with the use of real-time data. The real-time patient data are considered and taken from different systems to predict the options of treatment. The major part of healthcare involving ML is to identify diseases and diagnose the disease. Another application of ML involves the personalized medicine predictions using the data documented earlier. IBM Watson Oncology is one of the tools used for personalized medicine to track patient history and leverage treatment options. Artificial Intelligence and ML have a greater impact in predicting any kind of pandemic or diseases that arise [11].

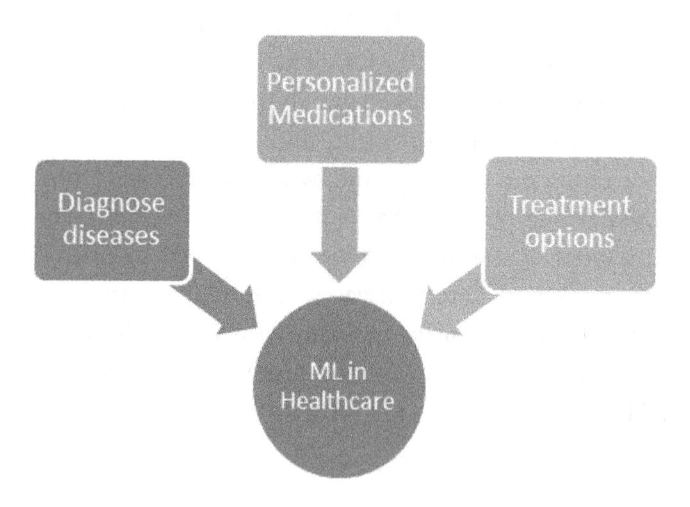

ML in Automobile Industry

ML helps car manufacturers to predict sales and even their manufacturing process. Vehicles use this concept to introduce driverless cars. Also it can assist drivers toward their journey. It is possible to predict and control the speed and movement of the cars, which can also direct and notify any kind of emergencies or dangers and prevent accidents. The usage of digital twin and ML helps in managing automobiles and making any kind of operations that can be handled and monitored virtually [12,13].

Implementation

There are quite a number of tools and packages for implementing ML algorithms. The packages include pandas, matplotlib, pycharm, Google Colab, scikit-learn, and keras.

Pandas

It is an open-source Python library used for data analysis. These are used for loading different file formats and manipulating the data. These packages help in carrying out the important processes of ML, like data preparation, cleaning, manipulation, modeling, and analysis.

Matplotlib

It is a 2-D python library used for plotting. It helps to visualize the model and derive conclusions out of it. It can be used along with all the different kinds of other packages for processing large amount of data.

Jupyter Notebook

This includes web applications that are very interactive and can share live code documents, and develop, execute, analyze, and discuss results. This has higher functionality and user scenarios.

It can also be used along with other tools and also supports different languages.

Google Colab

An interactive notebook that maintains the entire document and helps to share the code and execute along with other users. It is also similar to Jupyter Notebook. With Colab, we can make full use of the libraries for analyzing and visualizing the data. It also helps to train and test a large dataset and evaluate and make the predictions with visualizations. As it is collaborated with Google, it makes use of Google Servers and GPUS.

scikit-learn

It is the most commonly used Python library for ML. It consists of many different tools capable of implementing all learning algorithms and carrying out statistical modeling. scikit-learn has a lot of features, like supervised learning algorithms, unsupervised learning algorithms, and datasets and feature extraction. Installation of the packages like Pandas, Numpy, and Scipy before using scikit-learn is important.

Conclusion

Machine learning (ML) is a subset of artificial intelligence prevalent in almost all applications. ML is a process where computer systems mimic the human behavior and perform all activities faster than a human. In general, supervised learning algorithms can help to solve problems with lesser amount of data and can clearly label them.

Unsupervised learning gives better performance and results for large datasets. It is an ML technique, where you do not need to supervise the model. Unsupervised ML helps to find all kinds of unknown patterns in our datasets. These learning algorithms are used for unlabeled datasets. One of the main scopes of this algorithm is Quantum computers, which are good at manipulating high-dimensional vectors in large tensor product spaces. Development of both supervised and unsupervised quantum ML algorithms will increase the number of vectors and their dimensions exponentially than classical algorithms. With the rapid increase of technology and applications, ML takes a greater role in today's generation. It eventually makes human life easier in different aspects, such as disease predictions and monitoring, robots, driverless cars, agricultural field, and so on. As technology is growing, ML will also take up the world in a few years.

References

[1] https://www.digitalocean.com/community/tutorials/an-introduction-to-machine-learning#:~:text=Machine%20learning%20is%20a%20subfield%20of%20artificial%20intelligence%20(AI).&text=Because%20of%20this%2C%20machine%20learning,has%20benefitted%20from%20machine%20learning.

[2] https://medium.com/towards-artificial-intelligence/machine-learning-algorithms-for-beginners-with-ython-code-examples-ml-19c6afd60daa

[3] https://machinelearningmastery.com/a-tour-of-machine-learning-algorithms/

[4] https://monkeylearn.com/blog/classification-algorithms/
[5] https://www.machinelearningplus.com/predictive-modeling/how-naive-bayes-algorithm-works-with-example-and-full-code/
[6] https://livecodestream.dev/post/7-steps-of-machine-learning/
[7] https://towardsdatascience.com/introduction-to-machine-learning-for-beginners-eed6024fdb08
[8] https://towardsdatascience.com/types-of-machine-learning-algorithms-you-should-know-953a08248861#:~:text=Supervised%20learning%20algorithms%20try%20to,from%20the%20previous%20data%20sets.
[9] https://machinelearningmastery.com/types-of-classification-in-machine-learning/
[10] https://www.techradar.com/in/news/how-ai-and-machine-learning-our-improving-the-banking-experience
[11] https://technostacks.com/blog/machine-learning-in-agriculture/
[12] https://www.flatworldsolutions.com/healthcare/articles/top-10-applications-of-machine-learning-in-healthcare.php
[13] https://machinelearningmastery.com/supervised-and-unsupervised-machine-learning-algorithms

3

Next-Generation IoT Use Cases across Industry Verticals Using Machine Learning Algorithms

T. R. KALAIARASAN, SRUTHI ANAND, V. ANANDKUMAR, AND A. M. RATHEESHKUMAR

Department of Information Technology, Sri Krishna College of Engineering and Technology, Coimbatore, India

Contents

DOI: 10.1201/9781003119838-3

Introduction

Internet of Things (IoT) refers to the interconnection of devices embedded in our everyday life via the Internet. It also enables them to send and receive the data. The scope of the IoT is not limited to getting the devices connected/networked; it is much more than that as depicted in Figure 3.1. Aznd IoT comprises things that have unique identities and are connected to the Internet. IoT does not use the existing technology, rather a combination of knowledge base and experience is mandatory (Figures 3.1 and 3.2) [1].

The IoT workflow consists of four fundamental elements as shown in Figure 3.2:

1. Sensors
2. Connectivity
3. Data Processing
4. User Interface

Sensors

Sensors are nothing new; these are used by the organizations for a considerable period of time. Sensors play an important role in providing solutions using IoT. Sensors are devices that can detect the external environment. It is possible to collect data of almost every possible condition using sensors. There is a wide range of IoT sensors used to detect and measure various aspects.

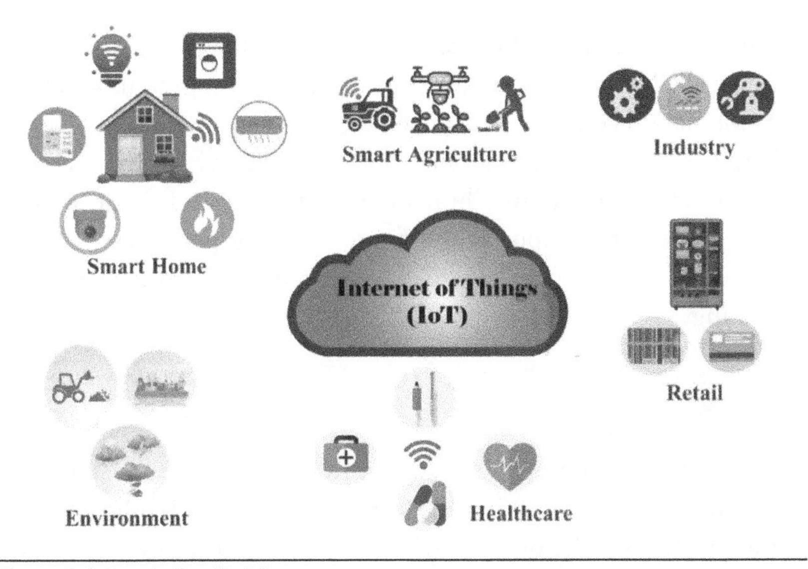

Figure 3.1 Domain-Specific IoT

Figure 3.2 IoT Workflow

Connectivity

IoT connectivity is defined as connection between devices such as sensors, routers, gateway, etc. There are a number of ways to connect to an IoT device, such as Bluetooth, Wi-Fi, Zigbee, NFC, Ethernet, LRWAN, etc.

Data Processing

Generally, sensors emit data that do not have any meaning (raw and unprocessed data). The information is inferred from the raw data by filtering and processing the data. Data processing involves input, process, and output.

Input: The data may be in the form of numbers, text, images, or videos. All these forms of data can be converted into machine understandable language.

Process: Different techniques to process data like classification, sorting, calculations, etc., are used to convert meaningful information from the data received.

Output: Although the information is inferred in the process phase, it is rendered into human understandable format in the output phase. The output in the form of text, graph, table, image, audio, video, etc., can be stored locally or in the cloud for further processing.

User Interface

A minimal user intervention is required for IoT where it is necessary for the user to understand the data captured by the system. A user interface is the point at which a user and a computer system can interact [2].

IoT in Industry

The major driving force behind Industry 4.0 is Internet of Things (IoT) [2]. IoT is used in the majority of industries, like manufacturing, logistics, oil and gas, aviation, and transportation. IoT 4.0 promotes and enhances automation, data collection, and analytics that can be useful for increased process optimization. It is observed that in 2016 for manufacturing operations in IoT, a total of $102.5 billion was spent. With respect to Industry 4.0, more focus in transportation is given toward freight monitoring. The evolution of IoT has enabled rethinking of almost all businesses to be moved toward the digital world. The devices connected via technology work independently to produce the output. The rapid growth of IoT has changed today's industries from factories to smart factories and smart production systems and so on. The data are collected through sensors, which contribute much toward identifying system failures for immediate maintenance. As the system can be integrated, it is used along with Machine Learning and Deep Learning systems to predict and recover from failures. These predictions contribute to better optimizations and gain insights yielding higher growth and productivity [3] (Figure 3.3).

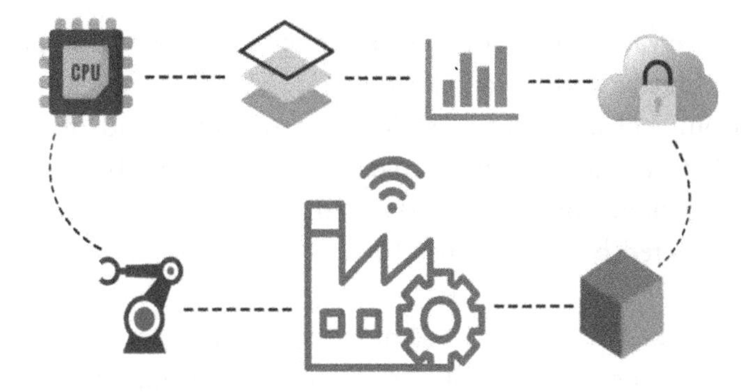

Figure 3.3 IoT in Industry

As shown in Figure 3.3, IoT enables the connected operation that has the ability to collect data from the connected devices, equipment, and sensors. The data can be processed, analyzed, and stored in the cloud for real-time process adjustments and optimization.

Machine Diagnosis and Prognosis System

Generally, mechanical equipment requires some manual maintenance, due to which there is an inability to diagnose equipment on time which sometimes leads to accidents. There is a number of mechanical equipments working from which the data can be collected in a short period of time. Using the collected data to improve the fault prediction becomes a challenging process [4].

Machine diagnosis refers to finding the cause of a fault in machine. Machine prognosis refers to predicting the performance of the machine. The data can be collected and analyzed on the basis of the performance of the predicted machine (Figure 3.4).

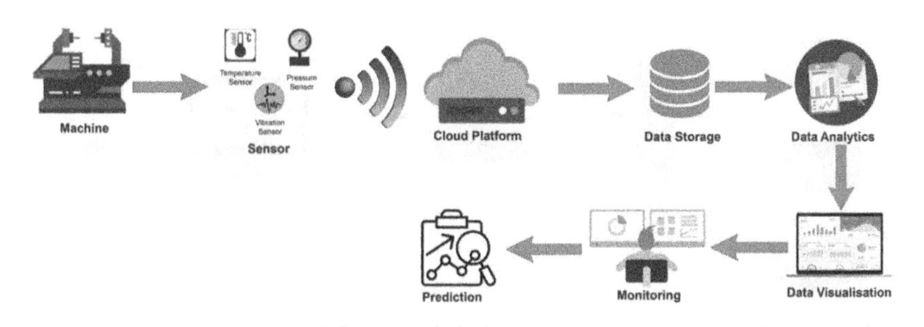

Figure 3.4 Machine Diagnosis and Prediction System

There are various types of sensors like temperature sensor, pressure sensor, vibration sensor, acceleration sensor, etc. Data can be collected from sensors on a timescale of a few milliseconds to seconds that leads to a huge amount of data. Those data can be sent to the cloud using cloud application and stored. The faults of certain devices can be detected by monitoring the health index of these devices in real time.

The data can be analyzed and monitored using various techniques to predict the performance of a machine by analyzing the data. The current operating conditions can be compared with normal operating conditions to find any deviations upon which the performance of a machine can be predicted.

Remote Monitoring and Production Control

There are many industries in production. Some of the industrial processes cannot be monitored by humans directly or near the production line. The industry may be small or large, some of the restricted areas need to be monitored. The most important feature of remote monitoring and production control in industry is the centralized supervision over the machines in the process of production. Information obtained from the devices like sensors, camera, etc. (Figure 3.5), provides a much clearer and faster insight into the actual production field. The staff is able to analyze the data with its assistance. This ensures that IoT is a pivotal technology in ensuring

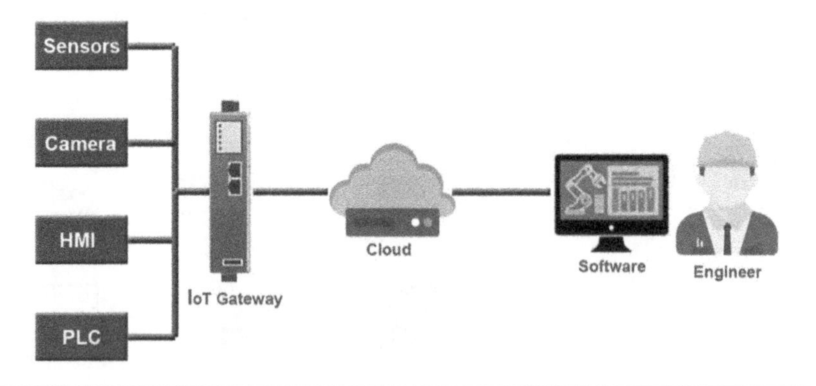

Figure 3.5 Remote Monitoring and Production Control System in Manufacturing Industry

production automation, tracking the status of the staff, and in monitoring the workers [5].

Indoor Air Quality Monitoring System

Quality of an air plays a vital role in industry. The quality of air determines the performance of workers that also affects operational cost and influences the productivity and efficiency of an industry [6].

The following are common pollutants that impact indoor air quality:

- Volatile organic compounds (VOCs)
- Carbon monoxide (CO)
- Particulate matter (PM)

Volatile organic compounds (VOCs): VOC is the main contributor to poor air quality. It emits gases from processes or products.

Carbon monoxide (CO): CO is a dangerous gas. It is impossible to see or smell. Sources of CO are automobile exhaust, generator, improper use of boilers, etc.

Particulate matter (PM): PM is a mixture of liquid and solid particles suspended in air, such as dust, pollen, smoke, and soot.

Different types of sensors are used to measure the air quality. Sensors are capable of detecting harmful toxic gases like carbon monoxide, nitrogen monoxide, nitrogen dioxide, particulate matter, and many more. Typically, the sensors can be placed across the industrial buildings rather than placing them on the ceiling or floor, wall mounting is the best suitable place to locate the sensors [7] (Figure 3.6).

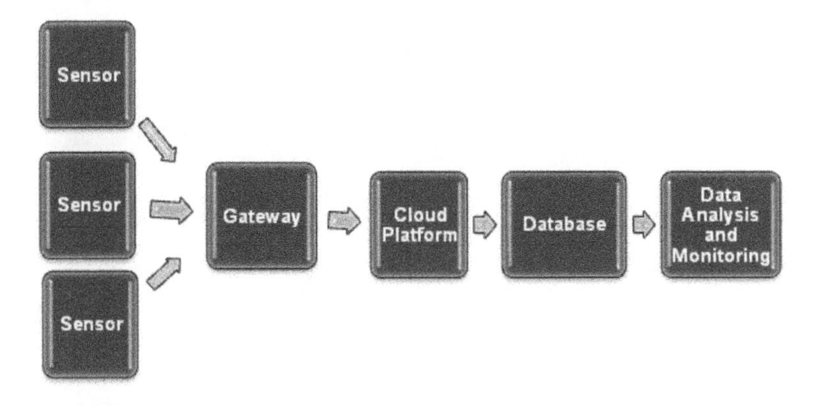

Figure 3.6 Indoor Air Quality Management System

Noise Monitoring System

Noise pollution is one of the major reasons for an impact on health. Noise generated by various industries like plants, mills, refineries, shipping docks, utilities, and more affects the surrounding environment, as well as workers, in the industry. By controlling and monitoring this noise, an industry can establish a working relationship which will benefit all [8].

This noise monitoring system helps in producing the noise map (Figure 3.7). The noise map is the graphical representation of the sound level of an environment. Number of noise monitoring stations is deployed at different locations of industry. Sound-level sensors used to monitor the data on the noise level from the noise monitoring station (Figure 3.8) will be collected and stored on server or cloud. The collected data can be analyzed and aggregated to generate the noise maps.

Figure 3.7 Noise Monitoring System

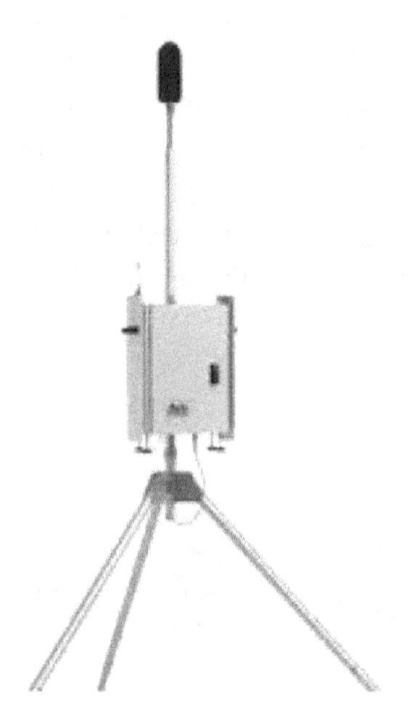

Figure 3.8 Noise Monitoring Station

Inventory Management System

Inventory management is a complex process specifically for large organizations. In this modern business environment, inventory is very important, as the uncertainty can happen at any point of time upstream or downstream. It is utmost important to keep the right level of inventory at the right time in the right place with right price and right time duration [9].

Inventory management using RFID tags will help the industry to maintain the right inventory level. RFID is based on radio wave technology to maintain and monitor the inventory. RFID fully differs from barcode. By using barcode, the products can be scanned and data can be returned, whereas in RFID, tracking can be done using RFID readers attached with warehouses or shelves. This will make them more powerful and useful for operators trying to find the stocks quickly and easily. IoT systems enable the remote monitoring of inventory using the data collected by using RFID readers.

Any materials or products can be tracked and fitted with RFID tags. There are two types of tags 1. Passive tag (Integrated chip and Antenna) 2. Active tag (Integrated Chip, Antenna and Battery). Each RFID tag is assigned with a Unique Identification Number (UIN). A RFID reader will extract UIN numbers and update in the database.

RFID reader sends energy to the tag for power. RFID tag sends data to the reader. And the reader will decode the data and send the data to the local server. A local server performs aggregation of data. An integration server will make the data available in enterprise resource planning (ERP) [10] (Figure 3.9).

IoT in Health Care

The role of IoT is increasing in the field of life sciences and medical health which is the need for today's society. It is explored for a number of purposes enabling end-to-end integration. The technology fosters the medical environment to connect patients and monitor them, which in fact improves systems as such. IoT can be integrated with any other technology to enhance health care systems starting from virtual online consultations to smart pills and wearables.

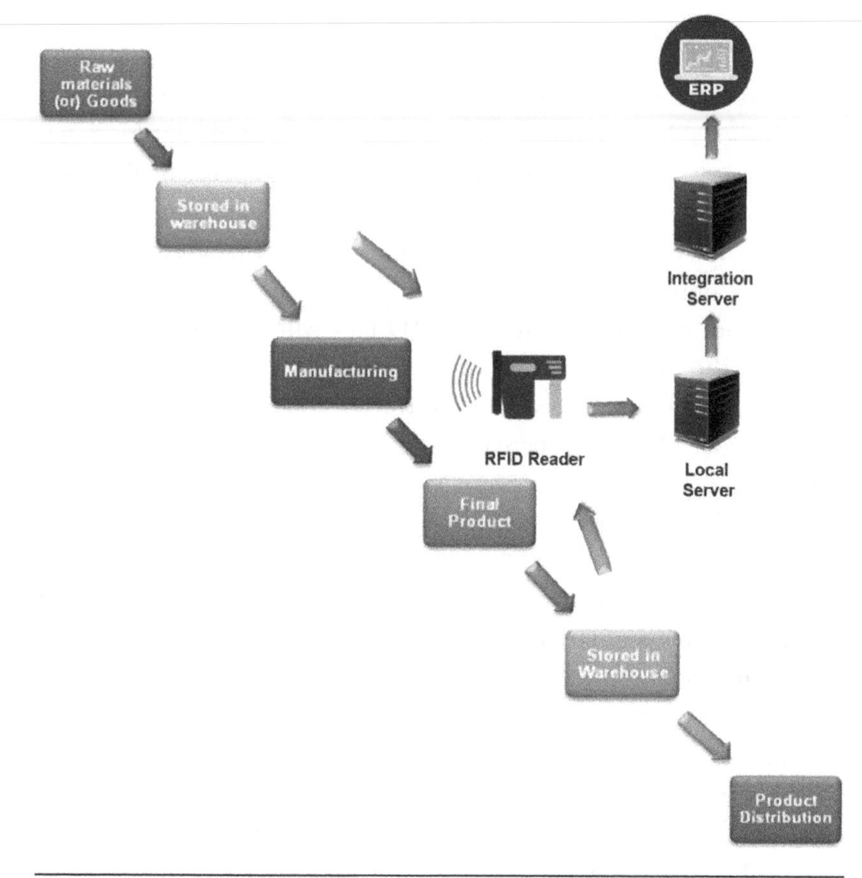

Figure 3.9 Inventory Management Using RFID

These technologies help people to lead a healthy life by keeping track of their lifestyle and eating habits that can enable the medical practitioners and patients to be connected through. Having a decentralized and dephysicalization kind of treatment is what is expected today. According to the reports by Statista, it is observed that there are about 26 billion connected IoT devices and by 2025, it is expected to grow to 75 billion. With respect to the global health care market, it is assumed that it will hit $543.3 billion per year. The current scenario in healthcare is the reactive sick care approach, where the patient is treated only after identification of the disease. But with the growth of IoT, it could be a proactive, predictive, or preventive approach.

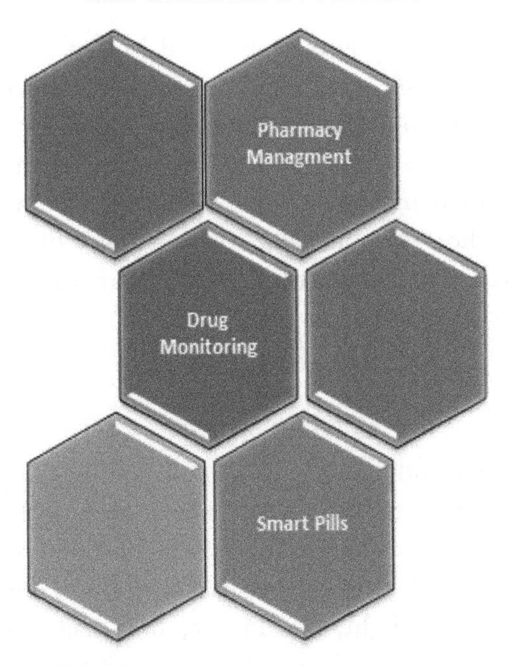

Figure 3.10 IoT in Health Care

The impact of IoT is that it can have a face-to-face communication to address the issues through the internet. This is extended up to providing health care advice and treatments even when they are on the way to the medical centers. It not only benefits the patient, but also notifies and informs the health officials to be prepared for the treatment prior to the patient reaching the hospital. Hence, it can be well understood that IoT plays a very vital role in health care [8]. There are different applications that contribute to IoT in health care, such as pharmacy management, smart pills, and drug discovery and management (Figure 3.10).

Pharmacy Management

In the field of drug manufacturing, it is very critical to make sure that there is no issue with the drug that is manufactured in terms of equipment and composition used in that drugs. This involves tracking the preparation of the drug process. There can be pitfalls due to the failure in the assets of the system, like mechanical damage of the equipment, lack of power supply, and lack of environment. To address these issues, IoT paves a greater way using sensors to monitor and track the working conditions of the equipment used for the manufacturing process. The machineries can be monitored for pressure, heat exchange, pH, compressors, and pumps. The data received from these sensors help to keep track of the machines, monitor the safety critical measures, and ensure safety of the workplace [11].

Drug Monitoring

Drug monitoring and preserving are one of the most important tasks in the field of medicines. It is well known that the drugs manufactured must be preserved in certain conditions concentrating on the environmental conditions, like temperature, humidity, light, carbon dioxide level, light, etc. Hence, IoT makes it possible to monitor the parameters and keep the pharmacists updated and alert them for any deviations. This will help to automatically control the leakage of any toxic substances and damage to the environment. Another function that IoT can perform in this field is to place sensors on the drug packages and monitor if it the package reaches the patients on time without any other serious issues. As it is known, there are issues arising due to the illegal transport of medicines and thus the quality may be compromised. Next is the issue in the process of supply chain management necessary to monitor the way in which the drugs are shipped. To ensure that this process is carried out, route details must be taken care of. This can include smart labels and tags used for packaging and GPS to continuously monitor how the vehicle travels. During shipment, it is also possible to keep track of the temperature and maintain ambience such that it would contribute to safely record and monitor the stability of the drug transported [12] (Figure 3.11).

The methods and techniques mentioned contribute to a greater extent in supply chain management. They also ensure that the patients or their attenders are informed about the conditions and details about the drug and how it is to be utilized. These are the advantages that IoT provides in terms of pharma application. Since these operations do not

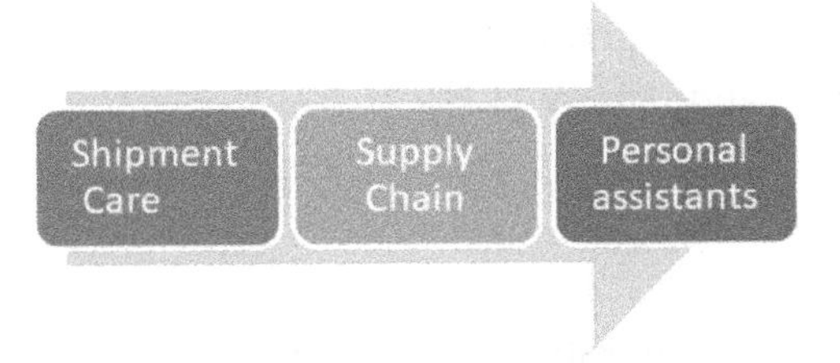

Figure 3.11 Drug Management

provide full transparency, it is difficult to ensure safety and security at times, which can also lead to discarding of drugs or patient mistreatment linking to some false electronic health records (EHR).

Smart Pills

Smart pills (Figure 3.12) are just like normal medicines monitored even after they have been taken. The term smart indicates that the pill is able to monitor the temperature, pressure, and keep track of the activities inside the body. It might sound similar to a pill cam also. This pill is used to treat the problems in the intestine, treat cancer, identify bacterial infection, etc. Most of the time, this pill continuously helps us to identify the issues in the intestinal region. There are a number of situations where a person may suffer from continuous vomiting or a kind of uneasiness in the stomach region. In this case, the smart pill enables the doctors to collect data as the pill travels through the body. This is done in order to identify the process that takes place when the food is swallowed. It does not take any kind of photos inside our human system but it monitors the movement. The smart pill is developed using either web or mobile

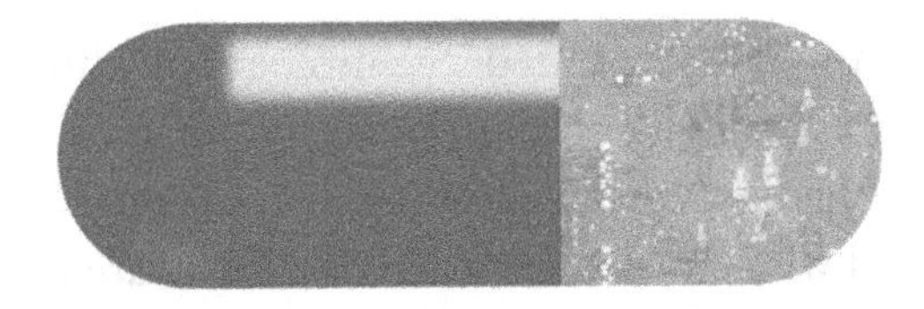

Figure 3.12 Smart Pills

applications that could track the process. Smart pills include camera, PCBs, antenna, and battery [13].

Wearables

IoT, as the name suggests, is the interconnection of devices, allowing the devices to communicate with each other by means of sensors. These sensors are connected to the bluetooth or their mobile applications, which help in analyzing and predicting the collected results. The sensors are embedded in wearables or as a smart device, or implanted onto some garments or fashion accessories or even imprinted as tattoos. Some examples of wearable technology include wristbands, watches, smart clothing, etc [14].

Fitness Trackers

Fitness trackers are smart devices for monitoring the activities of humans that include heart rate, breathing, pulse rate, and various body parameters. The fitness tracker measures the movements using a 3-D accelerometer. The data will be tracked every time the fitness tracker is worn on the body. These data are analyzed and shared with the device connected and displayed via a monitor. The movements are categorized in different factors and information is generated based on the different actions. The person makes use of the application and taps on it in which huge information is available and can be tracked [15].

Smart Hearing Aids

Hearing aids are much needed technology of this present era. These help to filter noise and sounds and manage hearing loss. These can easily adjust to the environment, thereby able to classify noise and sound without any disturbances. They are directly able to communicate to any of our smart device like TV or mobile phones without the need for any other intermediary devices. These devices not only help in rectifying or managing hearing impairment, but also in managing calls, stream audio, and easy understanding of the conversation. The main advantage is that these can also improve the quality of the sound received. These are usually placed at the back of our ears, which comprise a microphone, a sensor, and an amplifier to engage in the process. These aids are highly helpful for people who suffer from hearing loss due to ageing, a disease, or an injury [16].

Wearables to Monitor Diseases

Nowadays we come across a number of diseases that are not easily identifiable. Hence, these kinds of technology can help identify and alert in case of any abnormalities. It is highly used in detecting Alzheimers, Parkinsons, and Depression. Wearable ECG monitors help to detect pace, distance, elevation, and the different activities. In case our level increases, it can notify us by sending an alarm or a message. There are wearable blood pressure monitors who keep track of blood pressure and our daily activities. These are some of the common events. Apart from this, depression can also be controlled using IoT. It aids in identifying the mood of a person and also pacifies them both in normal and abnormal conditions. The solution could be the use of a Fitbit or a smart clothing that reduces their anxiety or disturbances to a greater extent (Figure 3.13).

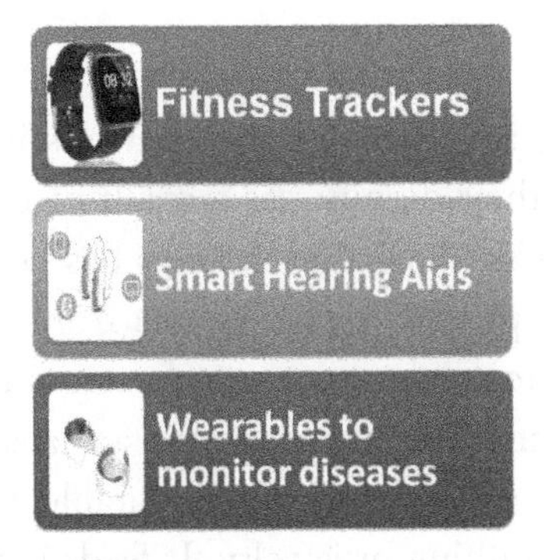

Figure 3.13 Wearables

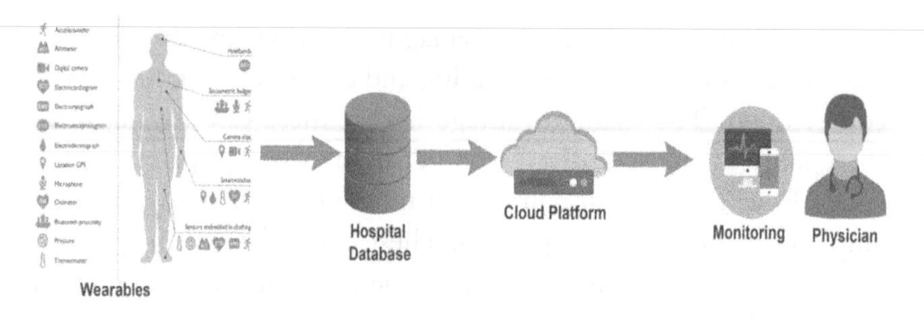

Figure 3.14 Monitoring Patient Using Wearables

Use of IoT for Physicians

Physicians benefit more with this technology. The data from the patient wearables help them to identify the treatment process, check if they are adhering to the treatments, and monitor them. It also promotes patient doctor interaction even in case of emergencies (Figure 3.14).

IoT sensors are being used in different medical equipments like nebulizers, wheelchairs, oxygen cylinders, glucose monitoring, and insulin detection. Wearable devices can transmit data directly to the practitioners that foster speedy diagnosis. It contributes to a greater extent in remote monitoring of patients even if the patient may not be able to communicate directly to the patients. Majority of the medical devices can be connected via this technology which may include wearables, labels that can be worn by patients or their attenders to reduce the admission process associated with it [17].

Medical Alert System

In this modern world, seeking assistance to aged persons is difficult. Many aged people fear losing their dignity. As a result of this, use of fall detection devices and other medical alert systems has been on the rise. Fall detection devices can prevent senior people from falling. In this device, a camera is placed to track senior people across the room. In addition to fall detection, medical alert systems are available for remote monitoring. This medical alert system is a wearable, which can be worn by an aged person as clothing or jewelry. It feeds into a series of connected sensors, which measure the health and welfare of the

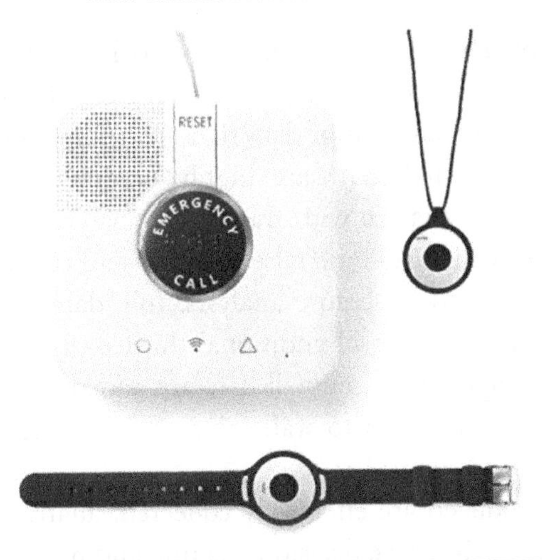

Figure 3.15 Medical Alert Device

person. If an aged person remains motionless for too long a period or falls out of bed, an alert is sent to the family members or friends who can help immediately [18] (Figure 3.15).

Data Analytics for IoT

IoT device will emit data, a raw data or an unprocessed data. Nothing can be done if only raw data are available. The information is inferred from the raw data by filtering and processing the data. From information, knowledge is extracted using organizing and structuring of information and decisions can be taken (Figure 3.16).

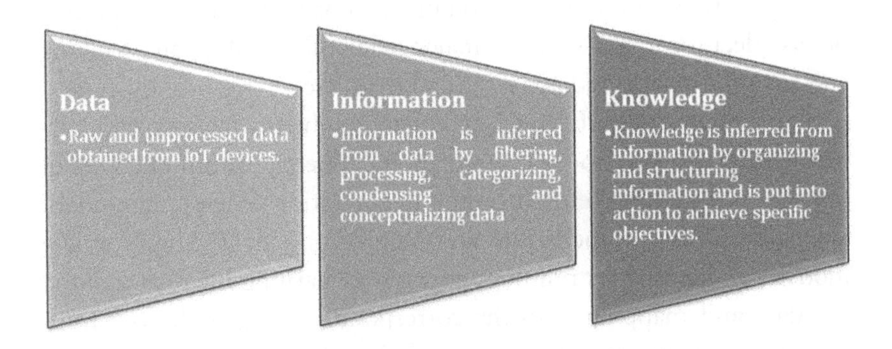

Figure 3.16 Data – Information – Knowledge

A variety of devices can be connected to Internet to share data. These data have no meaning without analysis. IoT data analytics is the analysis of huge amount of data that can be generated by various connected devices. IoT data are highly unstructured data, which make it difficult to analyze with traditional analytics. The data from the IoT devices will have some false readings and corrupted messages that must be cleaned up before analysis. IoT data are often meaningful only in the content of additional third-party data.

Example: In agriculture, rainfall data are added to help farmers so that they can decide when to water their crops along with moisture sensor data.

The data produced are effectively collected, analyzed, and stored. IoT data analytics improve decision-making and many more benefits, like improved equipment maintenance, operations optimization, and so on. The combination of IoT and data analytics has already been proven to be beneficial in various domains.

IoT Data Analytics Using Machine Learning Algorithms

With the emerging technologies in the field, IoT combines with cloud computing and data analytics for predictive maintenance and decision-making. IoT enables mechanical devices to be diagnosed and maintained with large amount of data collected instantly at different intervals of time. The data collected must be utilized effectively so that it increases the rate of fault prediction. The data received from the sensors are related to different equipments, and incorporating the concept of machine learning (ML) will enable efficient data analysis. ML algorithms tend to make good predictions that help in the process of easy decision-making across major fields, like health care, banking and financial sector, industrial automotives, etc.

The process of ML involves collecting huge datasets and implementing a model to train the datasets. The algorithm takes set of samples as a training set. With the categorization of learning models, viz supervised, unsupervised, and reinforcement, the appropriate models can be used for training. Supervised learning involves training the data and mapping it to the corresponding targets derived from the past experiences. Unsupervised learning techniques involve training the dataset without the need to supervise the data and it works on its own to draw conclusions. This is done by identifying patterns in the data and categorizing them, which is used in drawing conclusions. Having such models provide great insights in the field of IoT to model the data received from sensors. For instance, in the application of machine diagnosis, it is highly important to monitor the machine and also foresee if the machine needs some maintenance to meet the demands. Integrating IoT and ML generally has the following steps:

1. IoT Sensors are embedded to the devices.
2. Data from the sensors are loaded to the cloud.
3. Using an appropriate ML model, data are processed in the cloud using training and testing data separately.
4. The model evaluates on huge datasets to arrive at conclusions or come up with hypothesis.
5. This result is tested and validated.
6. This model is then evaluated on testing data and conclusions are drawn on the basis of the inferences that it has got from the trained data.

With these processes,companies benefit by applying them on different applications that help in giving detailed root cause analysis, fault predictions, maintenance routines, etc.

Conclusion

Internet of Things (IoT) is a technology in which every device is connected to some real-world objects. The basic idea behind this technology is the use of sensors to monitor and analyze data to make life more easy and comfortable. These can be used in almost every application from industry, health care, smart home, smart city, etc. This chapter focuses on the domains where IoT is prevalent in today's market. There are also certain issues with respect to using this with the expectation to be fully connected to a network every time and since every object is connected via the internet, it may be prone to security attacks. With this technology, there is no doubt that in the coming years, every organization may drastically move toward it. Hence, it requires the data to be available always and need to be stored for future retrieval and processing. It is the next step of an information age where it reduces human involvement and transforms it to machine controlled systems. The other important fact is that this technology will grow only if the computing power increases. With many numbers of devices connected, it is also essential to also care of the processing power and the network speed in which it can communicate.

References

[1] ArshdeepBahga and Vijay Madisetti, "*Internet of Things – A Hands - On Approach*", Universities Press, 2015, ISBN: 978-81-7371-954-7.
[2] Anandkumar, V., T. R. Kalaiarasan, and S. Balakrishnan. "IoT Based Soil Analysis and Irrigation System." *International Journal of Pure and Applied Mathematics* 119, no. 12 (2018): 1127–1134.
[3] https://www.i-scoop.eu/internet-of-things-guide/industrial-internet-things-iiot-saving-costs-innovation/
[4] https://www.sciencedirect.com/science/article/pii/S1569190X19301145
[5] https://www.scnsoft.com/blog/iot-in-manufacturing
[6] Sruthi Anand, S. Lavanya, and T. Eswari, "A Smart Monitoring System to Sense Short Circuit and Gas Leakage Using IOT", *International Journal of Advance Research in Computer Science and Management Studies*, vol. 5, no. 7, (July 2017), 1–6.
[7] https://www.iotone.com/usecase/indoor-air-quality-monitoring-iaq/u19
[8] http://www.larsondavis.com/applications/environmentalnoisemonitoring/industrialnoise
[9] N. Susila Dr., Sruthi Anand, Dr. Granty Regina Elwin J. Dr. Tsujatha, "Technology Enabled Smart Waste Collection and Management System Using IoT", *International Journal of Pure and Applied Mathematics*, vol. 119, no. 12 (2018), 1283–1295.
[10] https://www.peerbits.com/blog/warehouse-smart-inventory-management-solution.html
[11] https://www.ns-healthcare.com/analysis/iot-in-healthcare/
[12] https://www.pharma-iq.com/manufacturing/articles/the-role-of-iot-in-pharma-manufacturing-and-distribution
[13] https://www.iottechtrends.com/how-smart-pills-change-healthcare/
[14] https://www.businessinsider.in/science/latest-trends-in-medical-monitoring-devices-and-wearable-health-technology/articleshow/70295772.cms
[15] http://iotlineup.com/device/sync_smartband
[16] https://www.sdglobaltech.com/blog/10-brilliant-examples-of-wearables-in-healthcare
[17] https://www.finoit.com/blog/the-role-of-iot-in-healthcare-space/
[18] https://www.wirelesswatchdogs.com/blog/iot-applications-in-healthcare

4

A Panoramic View of Cyber Attack Detection and Prevention Using Machine Learning and Deep Learning Approaches

ESTHER DANIEL[1], N. SUSILA[2], AND S. DURGA[1]

[1]*Assistant Professor, Department of Computer Science and Engineering, Karunya Institute of Technology and Sciences, Coimbatore, India*

[2]*Professor, Department of Information Technology, Sri Krishna College of Engineering and Technology, Coimbatore, India*

Contents

DOI: 10.1201/9781003119838-4

Introduction

The malware is the code that performs malicious actions, such as causing computer operations disruptions, stealing sensitive information, accessing unauthorizedly, spying, flooding with spam, causing distributed denial of service attacks, and holding files for ransom. A concise glance at the historical backdrop of malevolent programming advises us that the presence of malware dangers has been with us since the beginning of computing. The most punctual archived infection showed up during the 1970s. It was known as the Creeper Worm and was a trial self-recreating program that replicated itself far off frameworks and showed the message: "I'm the creeper, catch me if you can." Afterward, in the mid-1980s, Elk Cloner showed up, a boot-division infection that focused on Apply II PCs. From these basic beginnings, a gigantic industry was conceived and, from that point forward, the battle against malware has never halted [1].

On looking into the statistics of the total number of malware since 2011 to date, we can see a drastic rise in the number of malware created. Due to the enormous rise in malware, the need for malware discovery has extended radically with the end goal that 5,638,828 distinct hosts were assaulted in 2018, according to the report given by Kaspersky labs. In a further article by Juniper Research, it was realized that more than 33 billion records will be taken by digital criminals in 2023 alone. According to Symantec, mobile malware variants increased by 54% in 2017, while up to 600% in IoT [2].

The Trojan malware was the most common type of malware. Different types of malware [3,4] can affect our devices and hinder privacy and performance.

1. **Trojans:** This malware camouflage itself as a legitimate app and gets installed on a device. This app can steal sensitive information, spy on activity, gain access to the device, and download other malware.
2. **Keyloggers:** This malware is designed to record or capture keystrokes when typed in an android device. These apps are readily available on the internet. Most of the apps masquerade

as parental control apps and some developers promote these apps as means to surveillance on other phones.

3. **Ransomware:** Ransomware is mostly for computer-based attacks, still, Ransomware affects android devices also. Ransomware encrypts all the data in a device and demands payment to decrypt the files.
4. **Spyware:** Spyware can be installed as a root kit or as an app, which monitors all the data including messages, phone calls, photos, videos, and location, and it can even use the camera for live streaming.
5. **Riskware:** It is an app that is not classified as malware, but as a potentially unwanted app. Riskware utilizes system resources in an undesirable manner and also poses security risks.
6. **Adware:** Adware steals your cookies information and gives advertisements based on it. Adware shows ads even when the device is not connected to the internet or when not using any apps. Adwares are considered as potentially unwanted apps, which get downloaded from third-party sites.
7. **Viruses:** Viruses wreck or even alter your touchy information present in the framework, which bombs the CIA triad.
8. **Root kit:** The assailant can do whatever he needs if he has a root kit infused in your framework. It stays disguised and is exceptionally hard to distinguish.
9. **Backdoors:** If backdoors are introduced, the interloper will never require any sort of confirmations to get into your framework. These can infuse infection, worms, and Trojan ponies when indirect access is introduced.

Given below are the percentages of various malware attacks on Windows systems.

Trojans are the most used malware to attack a windows system. When compared with other malware, Trojan attack percentage is around 64%, which is a big score. Virus is the second most used malware attack taking 15% when compared with all other malwares. Worms come in third place with 7% of attacks. These are the top three attacks that we have to be worried about. Ransomware is a deadly malware attack but when comes to the attack percentage, it has 0.93%, which is very less as compared with the top three.

The traditional method of identifying malware depends on signature-based malware detection. But nowadays, the malware attacks that occur are mostly zero-day malware, which increases the difficulty in detecting and classifying malware. Zero-day malware is a specific kind of malware or malicious software that has only recently been discovered where we wouldn't be able to classify or detect these malware [5]. AI comes into the picture and gives hope such that we are still able to detect and classify these types of malware. Various machine learning and deep learning algorithms are used to detect these types of malware and hopefully give a better result.

State of the Art Techniques

Android-Based Malware Detection

The android malware gained popularity and cybercrime kept increasing in the android environment. Most of the malwares began to be sold on the dark web, which is spread in the android community affecting thousands of devices. The first Android OS was launched back in 2008, ever since android OS has been a popular choice for smartphones. Now, 2.5 billion active Android users are present around the world. Google made the Android OS an open source for the smartphone manufacturers, thus the OS was made more customizable as per the manufacturers' need. Popular android manufacturers, such as Samsung, One Plus, Xiaomi, Oppo, Huawei, etc., made their skin for android OS and applied it over the OS architecture. Stock Android (i.e., The original OS) has been used by Google Pixels, Nokia, and some manufacturers. The use of skin over the stock OS ensured better performance specifications for the smartphone model; however, some manufacturers violated the privacy concerns of the users by collecting user statics data without proper concern from the users.

The first malware for android was reported in 2010, 'DroidSMS' a Trojan, an SMS fraud app. The app made automatic subscriptions on the affected phones. The process of subscription was made in the background without the users knowing about it. The same year, another Trojan 'TapSnake' was released. The app leaked the GPS location of devices over HTTP, making the location of the devices known by other devices through that GPS Spy app.

Classification of detection techniques based on the deep learning methods implemented for android malware detection over the years are highlighted briefly.

Mohammed K. Alzaylaee et al. [6] in their paper, "DL-Droid: Deep learning-based android malware detection using real devices," propose a deep learning system that uses dynamic analysis along with stateful input generation for detecting malicious Android applications. The experiment used 30,000 applications composed of benign and malware, which were presented on real devices. The study analyzed the code coverage and also compared the detection performance of the models used. The study produced a 97.8% detection rate for dynamic features and 99.6% for dynamic and static features, and the study achieved an accuracy of 98.5%.

Deep android malware detection by McLaughlin et al. [7] implements a convolution neural network for malware detection on android devices. The method was implemented by performing static analysis on the raw opcode sequence of the malware. The opcodes were obtained from the dissembler program and the malware indicative features were learned accordingly from the opcodes. The operation for training the network takes is easier to approach than the other n-gram methods used before, as this implementation removes explicit enumeration of millions of n-grams for training. The network structure allows the usage of long n-gram features, which is not feasible with other methods.

S. Hou et al., in their work "Deep4MalDroid," use a method known as Component Traversal, which automatically executes the app code routine for android completely. A weighted graph has been derived from the extracted Linux kernel system calls and a deep learning framework has been used to identify a new malware app. This method outperforms signature-based detection techniques which were used previously as anti-malware app [8].

DroidDeepLearner [9] uses deep learning to characterize and identify malware based on behavioral patterns. The proposed method emphasize on autonomous learning and solve issues without human intervention. The method uses both dangerous application programming interface (API) and risky permission to build the Deep-Belief Network (DBN) model. The experimental results have shown better accuracy than Support Vector Machine (SVM)-based solutions.

Droid-sec by Zhenlong Yuan et al. emphasizes on ML-based detection techniques utilizing 200 extracted features from static and dynamic analysis for android malware detection. From the study, it is observed that the deep learning techniques outperform machine learning techniques. The study produced a deep learning method for malware detection with 96% accuracy [10].

E. B. Karbab et al. propose MalDozer, [11] a convolution neural network (CNN)-based malware detection system for android. The proposed method uses less preprocessing for the assembly process and thus allows for quick preprocessing. MalDozer automatically extracts features starting from the API calls and learns malicious and benign patterns. The study used 33,000 malware apps and 38,000 benign apps for evaluation of the model and the results showed an F1-score of 96% with a false positive of 0.06%–2%.

Deep Flow analyzes the data flow for malware detection. The study extracted 323 features from android apps which included malicious and benign apps [12]. The study applied the SUSI technique to categorize data flow features combined with DBN deep learning model. The study achieved a higher F1-score and outperformed traditional machine learning models.

Hou et al., in their paper, propose DroidDelver for deep learning-based android malware detection. In this paper, features are represented as statically extracted API calls from smali codes, which are converted into code blocks. The proposed deep learning technique uses DBN to analyze generated code blocks from API calls. The study proposed an accuracy of 96.66% and out-performed other machine learning and deep learning models [13]. DroidDelver to detect the malware contains five major components: (1) Unzipper and decompile are used to unzip the APK files and decompile the dex files to smali codes; (2) API call extractors are used to extract the API calls from the smali codes; (3) API call block generators are used to extract the API calls and the smali codes are used to generate the blocks; (4) Deep Belief Network classifier is used as malware detection for API call blocks; and (5) Malware detector performs the above four components process.

The three levels of penetration technique for installation, activation, and running of malware on the android system are as follows: (1) repackaging is the type used to install the malicious thing on the android platform [14]. (2) Update is difficult to detect but downloads the malicious activity at runtime. (3) Downloading is a traditional attack technique that is needed by the users who download apps very keenly. The application features, API calls, and permissions are used in the framework to detect. The first step is an app analyzer that decompresses the apk files of an app and extracts the class files and AndroidManifest.xml. The second step is characterizing the app based on its permissions and API calls. The third step is used for feature extraction based on the Android Manifest and API calls from the class files. By using a support vector machine, decision trees, and bagging predictor's classifiers the Apps can easily be classified as benign or malware.

There exist different types of attacks on the android system [15]. The first step uses the static and dynamic analysis with code to insert the malware application in apps. The second step uses the modification of the system to access the changes occurring on the system. The third step involves virtualization to separate the domains of an application to install the multiple instances of the android system on the same device.

Ransomware Malware Detection

Ransomware is a type of malware that the attacker uses to extort money from the victim, by encrypting the files in the victim's system and holding them "hostage," until the victim provides the attacker with the demanded money. The attacker essentially blocks the victim's access to the files or will threaten the victim with the exposure of the files.

Ransomware can enter a system or network in several ways; the most common method is through phishing emails or links. Once a download is done from these spam emails or phishing links, the downloaded file then executes the Ransomware throughout the system. This malware is capable of spreading through a network in which the infected system is connected, thus effectively propagating through all the systems in the network and infecting them all. More sophisticated Ransomware is capable of spreading and working by itself, without human activity. Such Ransomware attacks are popularly known as drive-by attacks. In the past few years, the number of Ransomware attacks has reached an alarming rate.

According to Purplesec, a cyber security services company based in Washington, the number of new strains of Ransomware increased by 46% and the number of attacks by 41% in 2019 [16]. Mostly, budding or small businesses fall victim to this kind of attack, as they usually can't afford to invest in security. And the rise in Ransomware attacks can be owed to the fact that Ransomware services are provided on the dark web and other unsavory parts of the internet for a minimal price, and it requires almost no amount of technical knowledge to deploy. Therefore, it has become of the utmost importance to protect our systems from these kinds of malicious attacks by identifying and detecting the attacks. Here, we look at the various deep learning algorithms employed in detecting the Ransomware and conduct a thorough study to determine the pros and cons of each algorithm concerning the detection and identification of these attacks.

Smith Maniath et al. [17] has suggested the use of long-short term memory for identifying Ransomware by analyzing the behaviors of API calls. They suggest that analyzing the API call sequence of a process can explain the behavioral patterns of that process. Ransomware families have some common traits, such as short-term connection to Command and Control Center, deletion of Shadow Volumes, and employement of a huge amount of file system operations. By analyzing the API calls dynamically with the use of LSTM, a deep learning method, we can classify a binary as Ransomware. They have created a model of three layers that has 64 LSTM nodes in each layer and gained the highest training accuracy of 99.9%. While testing, they have achieved the highest accuracy of 96.67%, when compared to the previous work of implementing deep learning in malware analysis.

A Deep Neural Network (DNN) model that detects Ransomware tweets, is proposed by R. Vinayakumar et al. [18]. DNN consists of basically three types of layers, an input layer, one or more hidden layers, and an output layer. They have analyzed almost 25 Ransomware families from 2010 to 2017, used this model to comb through the social media sites for malicious tweets, and classified them to the corresponding identified Ransomware family. This model can automatically and continuously analyze the online posts, thus giving enabling incident management with better resources and procedures to mitigate the threat. They have suggested a Ransomware triage method that consists of three steps, namely, data collection and preprocessing, optimal features extraction, and classification algorithm.

Text Encoding is aepplied at the data collection and preprocessing stage, Bag of Words and Keras Embedding are implemented in the second stage, and DNN in the third stage. The proposed model was able to give a higher accuracy rate of 78.9% than the predecessors and has an average error rate of 27.8%.

Seong II Bae et al. [19], have proposed a model that identifies Ransomware, malware, and benign files using machine learning algorithms. The API invocation method although popular is not suitable for three-class detection; therefore, a new model has been suggested. Input vectors are generated from the n-gram sequences derived from the processed API sequence. The input vectors are represented as 1 if n-gram appears in the n-gram sequence and 0 if it does not appear. The weights of each element are decided using CF-NCF (Class Frequency Non-Class Frequency). Mainly a combination of six machine learning algorithms is used to create the testing model i.e., Random Forest (RF), Logistic Regression (LR), Naïve Bayes (NB), SGD (Stochastic Gradient Descent), KNN (K-Nearest Neighbor), and Support Vector Machine (SVM). The proposed method has proved to have an accuracy rate of 98.65%, which is relatively higher than the previous results. And experimental results also have proven that CF-NCF has much better performance than TF-IDF for the detection of Ransomware.

An extensive review of the existing Ransomware detection techniques and has been provided and the Ransomware lifecycle and its properties have been mentioned [20]. In addition to it, the pros and cons of each detection technique along with a proposed detection model are also provided. As a result of their extensive review, they have concluded that regression-type algorithms were the most preferred for Ransomware detection method, and the most important evaluation metric is considered to be the accuracy rate and true positive rate. In the proposed technique, they have suggested the implementation of a hybrid technique, which is a combination of regression and rule-based algorithms.

A regression algorithm can build a prediction model based on the predictor-outcome relationship. And a rule-based model produces a rule for its prediction model. They have stated that the correct combination of algorithms can complement one another, thus leading the hybrid algorithms to outperform individual ones.

Jinsoo Hwang et al. [21], have proposed a two-layer Ransomware detection model that incorporates Markov Model in the first stage and Random Forest Model in the second stage. Initially, they analyze the API sequences at runtime and filter out 303 unique APIs that represent the general API list. They have then implemented Markov Model to these API sequences to filter out the Ransomware calls. This gave a moderate error rate of 0.9744 ± 0.0159, and a fairly good false-positive ratio but a poor false-negative ratio around 20%. To eliminate this disadvantage, the Random Forest Algorithm is applied in the second stage. The aim was to find a threshold value that provides a false-positive ratio below 5% and a false-negative ratio as low as possible, which concludes 0.2 as the threshold value. For the given threshold value, the two-stage model gives an accuracy of 97.28%, 4.83% FPR, and 1.47% FNR.

Netconverse [22] is a machine learning evaluation study for dynamic detection of Ransomware in Windows network. This model has three phases, data collection, feature extraction, and implementation of machine learning classifier. A total of six machine learning classifiers are implemented, namely, Bayes Network, Multilayer Perceptron, J48, K Nearest Neighbors, Random Forest, and Logistic Model Tree. Two classifiers, namely, KNN and LMT were found to have an increase in model creation but the rest of the algorithms took decreased time rate to process owing to the decrease in attribute numbers. The J48 algorithm proved to be the best algorithm among the six, with the highest detection rate of 97.1%, followed by the LMT classifier that has a detection rate of 96.8%.

Ibrar Yaqoob et al., in this paper, have thoroughly explored the various aspects of a Ransomware attack, such as the various types of common Ransomware, the rise, the challenges of Ransomware attacks, security and the aspects in IoT [23]. A brief yet exhaustive account of the latest and most innovative security development against Ransomware in the field of IoT is given. Several case studies are also included to alert the readers on the various security aspects and threats of IoT devices and the different ways in which Ransomware can attack and exploit these vulnerabilities. They have also included several countermeasures and security details to secure the IoT devices and several research challenges and lack of proper Ransomware identification and detection systems are also discussed.

Rakshit Agrawal etal, have proposed a model in this paper that involves the comparison between the LSTM and ARI-LSTM in detecting and identifying Ransomware threads [24]. Attended Recent Inputs (ARI) is an enhanced neural cell to involve learning from Ransomware sequences. This cell processes the sequences of input and at the same time analyzes the history of sequences also. Here the implementation of ARI with LSTM has been developed, in which the LSTM is incorporated with the ARI mechanism and the resulting neural network sequence is used for sequence learning in Ransomware. In the experimental results, it has been proven that ARI-LSTM provides a much better accuracy ratio than LSTM. ARI-LSTM has an accuracy rate of 0.97 when compared to the 0.87 of LSTM.

Hanqi Zhang etal., in this paper, incorporates the use of N-gram sequencing, Term Frequency–Inverse Document Frequency (TF-IDF), and a comparison of five classification vectors to find the best one for Ransomware detection and identification [25]. The TF-IDF is calculated for each N-gram sequence, which is transformed from opcode sequences of the Ransomware samples. The TF-IDF vectors are calculated to select the featured N-gram so that they can exhibit better discrimination among the Ransomware families.

These vectors are then fed to the classification Machine Learning Classification Algorithms, namely, Decision Tree, Random Forest, K-Nearest Neighbor, Naïve Bayes, and Gradient Boosting Decision Tree. Among these algorithms, Random Forest has been proved to have the highest accuracy rate of 99.3%, while discriminating between Ransomware and trusted software and a recall of 99.8%. This method has also gained a 91.43% accuracy rate in detecting wannacry malware.

Windows-Based Malware Detection

The most recognized malware discovery techniques can be arranged into three principal strategies:

1. *Static Analysis:* It extract features from the file's structure without running the file.
2. *Dynamic Analysis:* The file is made to run and the features are extracted while the file is running.
3. *Hybrid Analysis*: This is essentially the mixture of both dynamic and static analyses.

In the further segments, we will be glancing through different ML and DL algorithms that had the option to identify and classify malware with the features acquired from the investigation stage.

Ihab Shahadat et al. [26], created various Machine Learning models for prediction, which includes both binary classification and multi-classification; for data splitting, cross-validation is used for the feature selection which minimizes the features. The feature is extracted using a heuristic strategy which is nothing but dynamic analysis. They have attained an accuracy of 98.2% in the Decision tree for binary classification and 95.8% in Random forest binary classification.

These authors [27] gathered the dataset obtained from both static and dynamic analysis required and stored it in Apache Spark. The data stored are pre-processed and the necessary features were extracted and sent to five diverse base classifiers. At that point the outcome is utilized to plan two distinctive outfit learning plans, for example, Weighted Voting and Stacking, which give a decent result. Big Data is utilized to deal with the malware dataset. We have utilized a fair dataset in this model. This model gives an exactness of 99.5% utilizing WeightedVoting_RACA.

This model build [28] firstly converts the binary malware file into a Gray-scale picture and utilized Visual-AT ML visualization to classify malware. It deals with adversarial examples and gives a superior precision rate. Adversarial examples are taken care of in this model. Visual-At gives an exactness of about 97.73%.

These researchers have fabricated a multi-modular profound learning system known as HYDRA that comprises four fundamental segments: API-Based parts, Mnemonics-based segments, Byte-based segments, and the element combination and characterization component. The impediments of end-end learning have been limited utilizing this method. The double substance can likewise be included with the current model which may give a superior result. About 99.75% is the precision identified while actualizing Multi-modal Deep Neural Network [29].

MalFCS [30] primarily has three stages. The double malware documents are changed over into entropy graphs, and the highlights are removed from the entropy graphs which at that point went into the SVM classifier for malware characterization. It utilizes entropy graphs instead of a grayscale picture. It can utilize GAN to create great manufactured entropy graphs and lift model execution. It can acquire astounding characterization execution with the exactness of 0.997 and 1 individually on the Mailing dataset and Microsoft dataset.

The model developers fabricated a model known as IMCFN which can be predominantly isolated into two sections as malware picture age where the double malware records are changed over into picture and CNN calibrating utilizing back-proliferation procedure which predicts whether the created picture is malware. When contrasted and CNN, IMCFN gives a superior result. The trial investigation demonstrated IMCFN gives an exactness of 97.56% [31].

A feature-hybrid malware variants detection using CNN-based opcode embedding and BPNN based API embedding model comprises of two principal sub-divisions [32], in the first division the Opcode is created from.exe record, and afterward utilizing an n-gram model an opcode n-gram is produced. At that point, PCA is utilized to lessen measurement and afterward went into a concealed layer and Softmax classifier is utilized to characterize. In the subsequent division, The API calls are produced from the.exe document and a vector recurrence is utilized to change API calls into vectors and the PCA is executed to choose the most required highlights and is sent into the neural network for preparation. At that point, a classifier is utilized to group malware. This is a component half and half malware variation model. It can include more static elements. This methodology achieves an exact pace of 95%.

Lu et al.'s model [33] utilizes the malware dataset and is imported in the cuckoo sandbox. They arranged an API call sequence utilizing a bidirectional LSTM model. Programming interface call affiliation and recurrence are examined utilizing the AMHARA algorithm. At that point, RF and LSTM are joined and anticipated, which give a superior outcome. ML and DL are both consolidated in this model. Only a little dataset is utilized to prepare the model. The exactness gave in this joined model is 96.7%.

Suyeon Yoo et al., proposed a half and half model under the mix of Random Forest and Deep Learning (MLP) utilizing 12 hidden layers to distinguish whether the document is malware or kind [34]. The discovery depends on casting a ballot on the output gave by these two models. It has high TP and low FP, which are significant in Malware Detection. The accuracy in this model is just 85.1% which can be improved. It provides a TP pace of 85.7% and FP pace of 16.1% with a presentation of 60.9s per record.

A MalDy structure [35] is built which initially collects the behavioral report from sand-boxing and creation runtime and convert those report into a BoW utilizing NLP. At that point, they utilize different AI models to classify. Right off the bat, there is a dangerous location model and n number of models for n no. of malware family order. This structure is versatile in any environment. The model is reliant on the execution condition report system. MalDy accomplishes over 94.1% of F-score in android and accomplishes a precision pace of 94% on Win32.

Built-up a lot of regression models to separate the connections between the highlights of cutting edge malware known as Stuxnet and afterward ML calculations are utilized to foresee malware dependent on those features. This model can anticipate progressed malware as same as Stuxnet. Just features related to Stuxnet are utilized for expectation. At the point when utilized RF the outcome has $R2 = 0.8203$ for Stuxnet malware [36].

Linux-Based Malware Detection

Linux-based Malware or malicious software is used to harm the host operating system or to steal the sensitive data of the user, organization, and companies. It helps to gather the information illegally. Malware detection is used to detect the presence of malware on a host operating system or used to detect any malicious activity in the system. Malware detection is mandatory for the system to not lose the sensitive information of any user. Malware detection is performed using different techniques but in this survey, we are using machine learning algorithms to detect the malware on Linux operating systems. Nowadays malware is a great problem for personal data. Many approaches are used to identify the malicious executable linkable files (ELF). Linux operating system is open-source software and released under General Public License(GNL).

The three-fold approach for the proposed method is: (1). logging of system call traces; (2). methods for the elimination of redundant features; and (3). Usage of prediction models [37]. This method is proposed to give good computational efficiency without losing accuracy. Kruskal–Wallis test is a rank test based on the differences of independent medians. It is used to describe whether the two classes have an equal median or not. If the two classes have an equal median, then it is not in classification. Deviation From Poisson (DFP) is used to select the robust vector for classification. The parameters of the system are accuracy and feature length.

The malware sample dataset is taken from VX-Heavens and benign samples are taken from directories of Linux which are in Executable linkable files format [38]. System call logging is executed using the Linux platform "strace." By using arguments and return values extraction of system calls is done after this preprocessing of system calls is done for features. After the system call extraction, the data set is divided into training level and testing level. The training step is used for classification models and the testing test is used for prediction models. After this step feature analysis and selection, the level is taken by two steps: preparing different feature categories and feature selection. Feature categories are union, intersection, discriminating features for malware, and discriminating features for benign. Feature selection is a class discrimination measure (CDM) that is used to improve the classification for multiple data sets, the main concept of this is the odd ratio and elimination of sparse features (ESF).

Machine learning systems are based on one-sided perceptron and a kernelized one-sided perceptron for the datasets to clean the data and the malware files [39] Gavrilut et al. framework takes three datasets: training dataset, testing dataset, and scale up dataset. Cross-validation is done to choose the correct values from the parameters. The parameters of this paper are feature selection measures and cross-validation with the cascade one-sided perceptron algorithm.

To produce security and privacy for mobile users, the parallel machine learning technique is used because it uses multiple algorithms in a single unit. Supervised algorithms like Decision Trees, Simple Logistics, Naive Bayes, and Part, Ridor are used to detect the malicious on the mobile platform [40]. API calls related, command related, and permissions are used to extract the datasets.

Malware Detection Tools Used

Malware such as are viruses, worms, Trojan horses, spyware, adware, and Ransomware were discussed. Malware is a threat to the system and spreads due to vulnerable or carelessness of users. The various tools used for malware detection are listed below [41–43].

TOOL USED	DESCRIPTION
PEiD	It is an instrument which distinguishes the greater part of the normal packers, cryptors, and compilers for a PE (Portable Executable) records.
HashCalc	HashCalc calculates many cryptographic hashes for your files.
Dependency Walker	Dependency Walker is used to list the imported and exported functions of a portable executable file. It also displays a recursive tree of all the dependencies of the executable file
CFF Explorer	CFF Explorer was designed to make PE editing as easy as possible, but without losing sight on the portable executable's internal structure
Resource Hacker	Resource Hacker is used to add, modify, or replace most resources within Windows binaries including strings, images, dialogues, menus, Version-info, and Manifest resources
DIE	Detect It Easy, or abbreviated "DIE," is a program for determining types of files.

(*Continued*)

TOOL USED	DESCRIPTION
rEMnux	rEMnux is a lightweight Linux distribution for assisting malware analysts in reverse-engineering malicious software.
JavaSnoop	JavaSnoop attempts to attach to an existing process (like a debugger) and instantly begin tampering with method calls, run custom code, or just watch what's happening on the system.
Valgrind	Valgrind is a programming tool for memory debugging, memory leak detection, and profiling.
Zeek	Zeek sits on a "sensor," hardware, software, virtual, or cloud platform that quietly and unobtrusively observes network traffic.
diStorm	diStorm is very robust and mature and used widely all over the world. It is also the fastest disassembler in the world and is still highly maintained and updated by its creator.
PEview	PEview is a viewer for PE files. PEview is a lightweight program, being a small standalone executable around 70 kb in size. For determining basic PE information
Bintext	A little, quick, and amazing book extractor that will be quite compelling to developers. It can separate content from any sort of record and incorporates the capacity to discover plain ASCII text, Unicode text, and Resource strings, giving helpful data to everything in the discretionary "progressed" see mode. Its extensive separating forestalls undesirable content being recorded
MD5deep	md5deep is a set of programs to process MD5,SHA-1, SHA-256, Tiger, or Whirlpool message digests on a discretionary number of documents.
IDA pro	IDA Pro is primarily a multi-platform, multi-processor disassembler that translates machine-executable code into assembly language source code to debug and reverse engineering. It can be used as a local or as a remote debugger on various platforms.
FileAlyzer	This is also an application that is used to access the PE file header but it is also equipped with a VirusTotTab to submit to the Virus total for analysis and functionality to unpack UPX and PECompact packed files.
Procman	Procmon, or Process Monitor, is a free tool developed by Windows SysInternals and is used to monitor the Windows file system, registry, and process activity in real time.
Process Explorer	Process Explorer is used to monitor the running processes and shows you which handles and DLLs are running and loaded for each process.
RegShot	Regshot is a great open-source utility to monitor your registry for changes by taking a snapshot that can be compared to the current state of your registry.
ApateDNS	ApateDNS is a tool for controlling DNS responses and acts as a DNS server on your local system. ApateDNS will spoof DNS responses to DNS requests generated by the malware to a specified IP address on UDP port 53.

For continuous live data feeds, the deep learning architectures such as generative adversarial network, shallow deep learning feature extraction, gradient boosting algorithm, Recurrent Neural Network (RNN), and DNN are used. The Convolution Neural Network (CNN)-based architectures are used for image classifications and analysis.

The deep learning architectures mimics the human brain thinking process and develops the design patterns for decision-making that improve the accuracy of malware detections and measures to prevent it. The various deep learning algorithms and architectures suitable for cyber applications were elaborated.

Conclusion

This chapter discussed the malware types found in real time specifically for android and Windows operating systems. The base for malware detection starts with static and dynamic analysis. Other than this, signature-based malware detection gives a better result. But nowadays, zero-day malware is drastically increasing which paves the way to machine learning and deep learning technologies. This work also illustrated the demand for real-time dynamic identification of malware instead of signature-based detection techniques. Various deep learning and machine learning techniques used for the detection of malware in android and other operating systems such as Windows and Linux were analyzed.

The significance and methods adopted in each technique were elaborated in-depth. Ransomware malware was studied in detail. From the state-of-art methods, it can be concluded that the combination of the algorithms, such as Random Forest, Logistic Regression, Naïve Bayes, Scholastic Gradient Descent, K-Nearest Neighbor, and Support Vector Machine, has yielded the highest accuracy rate (98.65%) in identifying and detecting Ransomware. The need for malware detection and classification has increased now in this technology era filled with the Internet. We have discussed the various tools utilized by machine learning and deep learning techniques for the detection and classification of various malware.

References

[1] Gilbert, Daniel Mateu, Carles, Planes, Jordi, "The rise of machine learning for detection and classification of malware: Research developments, trends and challenges", *Journal of Network and Computer Applications*, Vol. 153, 2020, pp. 1–22, ISSN 1084-8045.

[2] Symantec 2018 Internet Security Threat Report. Tech. rep. Symantec Corporation, https://www.symantec.com/content/dam/symantec/docs/reports/istr-_23-_executive-_summary-_en.pdf.

[3] Singh, A., Handa, A., Kumar, N., and Shukla, S. K., Malware classification using image representation, in: *International Symposium on Cyber Security Cryptography and Machine Learning*, Springer. pp. 75–92, 2019.

[4] https://www.geeksforgeeks.org/malware-and-its-types

[5] https://www.techopedia.com/definition/29741/zero-day-malware

[6] Alzaylaee, Mohammed K., Yerima, Suleiman Y., Sezer, Sakir, DL-Droid: Deep learning based android malware detection using real devices, *Computers & Security*, Vol. 89, 2020, 101663, ISSN 0167-4048, https://doi.org/10.1016/j.cose.2019.101663.

[7] McLaughlin, Niall, et al. "Deep Android malware detection," *Proceedings of the Seventh ACM on Conference on Data and Application Security and Privacy*. 2017.

[8] Hou, S., Saas, A., Chen, L., and Y. Ye, "Deep4MalDroid: A Deep Learning Framework for Android Malware Detection Based on Linux Kernel System Call Graphs," *2016 IEEE/WIC/ACM International Conference on Web Intelligence Workshops (WIW)*, Omaha, NE, 2016, pp. 104–111, doi: 10.1109/WIW.2016.040.

[9] Wang, Z, Cai, J., Cheng, S., and Li, W., "DroidDeepLearner: Identifying Android malware using deep learning," *2016 IEEE 37th Sarnoff Symposium*, Newark, NJ, 2016, pp. 160–165, doi: 10.1109/SARNOF.2016.7846747.

[10] Yuan, Zhenlong, et al. "Droid-sec: Deep learning in android malware detection," *Proceedings of the 2014 ACM conference on SIGCOMM*. 2014.

[11] Karbab, E. B., Debbabi, M., Derhab, A., and Mouheb, D., "MalDozer: Automatic framework for android malware detection using deep learning," *Digital Investigation*, vol. 24, 2018, pp. S48–S59.

[12] Zhu, D., Jin, H., Yang, Y., Wu, D., and Chen W., "DeepFlow: Deep learning-based malware detection by mining Android application for abnormal usage of sensitive data," *2017 IEEE Symposium on Computers and Communications (ISCC)*, 2017.

[13] Hou, S., Saas, A., Ye, Y., and Chen, L., "DroidDelver: An Android Malware detection system using deep belief network based on API call blocks." *Web-Age Information Management Lecture Notes in Computer Science*, 2016, vol. 9998, pp. 54–66.

[14] Peiravian, N., and Zhu, X., "Machine learning for Android malware detection using permission and API calls," *2013 IEEE 25th*

International Conference on Tools with Artificial Intelligence, Herndon, VA, 2013, pp. 300–305, doi: 10.1109/ICTAI.2013.53.

[15] Abikoye, Oluwakemi, Gyunka, Benjamin, and Oluwatobi, Akande, "Android Malware Detection through Machine Learning Techniques: A Review. *International Journal of Online and Biomedical Engineering (iJOE)*, vol. 16, 2020, p. 14. doi: 10.3991/ijoe.v16i02.11549.

[16] https://purplesec.us/resources/cyber-security-statistics/ransomware/

[17] Maniath, S., Ashok, A., Poornachandran, P., Sujadevi, V. G., Au, P. S., and Jan, S., Deep learning LSTM based ransomware detection. In 2017 Recent Developments in Control, Automation & Power Engineering (RDCAPE) (pp. 442–446). IEEE.

[18] Vinayakumar, R., Alazab, M., Jolfaei, A., Soman, K. P., and Poornachandran, P. (2019, May), "Ransomware triage using deep learning: Twitter as a case study." In *2019 Cybersecurity and Cyberforensics Conference (CCC)*, pp. 67–73. IEEE.

[19] Bae, S. I., Lee, G. B., and Im, E. G. (2020), "Ransomware detection using machine learning algorithms," *Concurrency and Computation: Practice and Experience*, vol. 32, no. 18, e5422.

[20] Kok, S., Abdullah, A., Jhanjhi, N., and Supramaniam, M. (2019), Ransomware, threat and detection techniques: A review. *Int. J. Computer Science and Network Security*, vol. 19, no. 2, 136.

[21] Hwang, J., Kim, J., Lee, S., and Kim, K. (2020). "Two-Stage Ransomware Detection Using Dynamic Analysis and Machine Learning Techniques." Wireless Personal Communications, vol. 112, no. 4, 2597–2609.

[22] Alhawi, O. M., Baldwin, J., and Dehghantanha, A. (2018). "Leveraging machine learning techniques for windows ransomware network traffic detection." In *Cyber Threat Intelligence* (pp. 93–106). Cham, Germany: Springer.

[23] Yaqoob, I., Ahmed, E., ur Rehman, M. H., Ahmed, A. I. A., Al-garadi, M. A., Imran, M., & Guizani, M. (2017). "The rise of Ransomware and emerging security challenges in the Internet of Things." *Computer Networks*, vol.129, 444–458.

[24] Agrawal, R., Stokes, J. W., Selvaraj, K., & Marinescu, M. (2019, May). "Attention in recurrent neural networks for ransomware detection." In *ICASSP 2019-2019 IEEE International Conference on Acoustics, Speech and Signal Processing (ICASSP)*, pp. 3222–3226, IEEE.

[25] Zhang, H., Xiao, X., Mercaldo, F., Ni, S., Martinelli, F., & Sangaiah, A. K. (2019). "Classification of Ransomware families with machine learning based on N-gram of opcodes." *Future Generation Computer Systems*, 90, 211–221.

[26] Shhadat, Ihab, Bataineh, Bara', Hayajneh, Amena, and Al-Sharif, Ziad A., "The use of machine learning techniques to advance the detection and classification of unknown malware." *Procedia Computer Science*, vol. 170, 2020, pp. 917–922, ISSN 1877-0509.

[27] Gupta, Deepak, and Rani, Rinkle, "Improving malware detection using big data and ensemble learning." *Computers & Electrical Engineering*, vol. 86, 2020, Article ID: 106729, ISSN 0045-7906.
[28] Liu, Xinbo, Lin, Yaping, Li, He, and Zhang, Jiliang, "A novel method for malware detection on ML-based visualization technique." *Computers & Security*, vol. 89, 2020, Article ID: 101682, ISSN 0167-4048.
[29] Gibert, Daniel, Mateu, Carles, and Jordi Planes, "HYDRA: A multimodal deep learning framework for malware classification." *Computers & Security*, vol. 95, 2020, 101873, ISSN 0167-4048.
[30] Xiao, Guoqing, Li, Jingning, Chen, Yuedan, and Li, Kenli, "MalFCS: An effective malware classification framework with automated feature extraction based on deep convolution neural networks." *Journal of Parallel and Distributed Computing*, vol. 141, 2020, pp. 49–58, ISSN 0743-7315.
[31] Vasan, Danish, Alazab, Mamoun, Wassan, Sobia, Naeem, Hamad, Safaei, Babak, and Zheng, Qin, "IMCFN: Image-based malware classification using fine-tuned convolution neural network architecture." *Computer Networks*, vol. 171, 2020, 107138, ISSN 1389-1286.
[32] Zhang, Jixin, Qin, Zheng, Yin, Hui, Ou, Lu, and Zhang, Kehuan, "A feature-hybrid malware variants detection using CNN based opcode embedding and BPNN based API embedding." *Computers & Security*, vol. 84, 2019, pp. 376–392, ISSN 0167-4048.
[33] Xiaofeng, Lu, Fangshuo, Jiang, Xiao, Zhou, Shengwei, Yi Jing, Sha, and Lio, Pietro, "ASSCA: API sequence and statistics features combined architecture for malware detection." *Computer Networks*, vol. 157, 2019, pp. 99–111, ISSN 1389-1286.
[34] Yoo, Suyeon, Kim, Sungjin, Kim, Seungjae, Byunghoon Kang, Brent, "AI-HydRa: Advanced hybrid approach using random forest and deep learning for malware classification," *Information Sciences*, vol. 546, 2021, pp. 420–435, ISSN 0020-0255.
[35] Karbab, ElMouatez Billah, and Debbabi, Mourad, "MalDy: Portable, data-driven malware detection using natural language processing and machine learning techniques on behavioral analysis reports," *Digital Investigation*, vol. 28, Supplement, 2019, pp. S77–S87, ISSN 1742-2876.
[36] Bahtiyar, Şerif, Yaman, Mehmet Barış, and Altıniğne, Can Yılmaz, "A multi-dimensional machine learning approach to predict advanced malware." *Computer Networks*, vol. 160, 2019, pp. 118–129, ISSN 1389-1286
[37] asmitha, K. A., and Vinod, P., "Linux malware detection using non-parametric statistical methods," *2014 International Conference on Advances in Computing, Communications and Informatics (ICACCI)*, New Delhi, 2014, pp. 356–361.
[38] Asmitha, K. A., and Vinod, P., "A machine learning approach for linux malware detection," 2014 International Conference on Issues and

Challenges in Intelligent Computing Techniques (ICICT), Ghaziabad, 2014, pp. 825–830, doi: 10.1109/ICICICT.2014.6781387.

[39] Gavriluţ, D., Cimpoeşu, M., Anton, D., and Ciortuz, L., "Malware detection using machine learning." *2009 International Multiconference on Computer Science and Information Technology*, Mragowo, 2009, pp. 735–741, doi: 10.1109/IMCSIT.2009.5352759.

[40] Muthuswamy, Sujithra (2017), "Detection of malware applications in mobile devices using supervised machine learning algorithms." *International Conference on Business Analytics and Intelligence*, IIM Bangalore, December 2017.

[41] Aslan, Ömer (2017), *Performance comparison of static malware analysis tools versus antivirus scanners to detect malware.* In International Multidisciplinary Studies Congress (IMSC).

[42] https://www.hackingtutorials.org/malware-analysis-tutorials/basic-malware-analysis-tools/

[43] https://www.hackingtutorials.org/malware-analysis-tutorials/dynamic-malware-analysis-tools/

5

REGRESSION ALGORITHMS IN MACHINE LEARNING

USHA SAKTHIVEL[1], NEHA SINGHAL[2], PETHURU RAJ CHELLIAH[3], AND ASHWINI R. MALIPATIL[4]

[1]*Dean Research and Innovation, Rajeswari College of Engineering, Bengaluru, India, Prof. & Head, Dept. of CSE, Visvesvaraya Technological University Belagavi, Karnataka, India*
[2]*Dept. of ISE, Sri Krishna College of Engineering and Technology, Coimbatore, India*
[3]*Chief Architect, RJIL, Bengaluru*
[4]*RRCE, Dept. of CSE, Sri Krishna College of Engineering and Technology, Coimbatore, India*

Contents

DOI: 10.1201/9781003119838-5

Introduction

In machine learning (ML), various kinds of algorithm are used for the prediction of results based on the provided data. The relationship that exists among the input data and the expected output is used for foretelling results based on the prediction. Regression algorithm is a prediction method that works on supervised learning techniques. These supervised algorithms allow machines to train and learn so as to include some kind of automation in outcomes. The predicted output is calculated on the basis of continuous numerical values in regression problems. Regression algorithms are categorized in linear and non-linear algorithms. Regression algorithms are used in finance forecasting, market trend analysis, time series prediction, and even for drug response modeling for COVID-19 situations. Regression algorithms use continuous numerical analysis on the patterns to perform the predictions of future outcomes. Some of the popular types of regression algorithms are simple liner regression, polynomial regression, support vector regression, decision tree regression, and random forest regression.

Background Study

Punal Pahwa et al. [1]. The paper works on the analysis of stock exchange marketing where the regression method is used as a strong predictor in the ML area. The regression method provides a statistics accuracy in the investment and stock exchange. Stock market fully depends on the prediction of the shares in the present and future. Hence, ML algorithm with the regression methods predicts the future stock price for the business.

Zhixia Yang et al. [2]. The paper works on multiclass classification in ML with the ordinal regression method. Multiclass classification works on the construction of decision tree. Support vector machines are used as tools for the classification. The paper represents the kernel ordinal regression for three class methods.

Ekta Ghambir et al. [3]. The paper work on the analysis and prediction of the spread of novel Corona virus, which helps in taking future precaution and necessary steps. With the implementation of ML model with polynomial regression algorithm, it was possible to predict the spread of virus in the next 60 days in India. Regression algorithm is a supervised ML algorithm, which reads the previous data, analyze it, and provide the accurate prediction on the live data. The extraction and selection of data need to be done for the best results from this model.

Shen Rong et al. [4]. The paper analyzes the iced product affected by the change in the temperature conditions. Linear regression model works on the relationship between the independent and dependent data. In this work, forecast temperature is taken as the independent variable and the iced product as the dependent variable. The implementation of leaner regression is done on the Python 3.6 platform with various packages. The prediction results given by the regression model help the company to have a control over the production and sale of the iced products in the market.

Ki-Young Lee et al. [5]. The paper works on the comparison of linear regression and artificial neural networks with the help of deep learning and linear regression algorithm through the supervised learning. AI works on two methods: supervised and unsupervised. Supervised is when the human provides the data and logical interference. Unsupervised is when the computer takes the data and performs the task with the supervision of the previewed data. The paper works to predict the future industry profit using the regression model using neural AI.

Khushbu Kumari et al. [6]. The paper explains about the analysis of linear regression model. Regression model is a statistical model that compares the relationship between the dependent and independent variables of the real-time dataset. Here, the patients having blood pressure are taken as the variable for the prediction of variation in the blood pressure with different variations in the patient body conditions. The linear regression model predicts the increase in the BP of the person with the change in his daily routine of his health condition.

Karunya Rathan et al. [7]. The paper speaks on the crypto currency, which is also called Bitcoin and is considered the asset of internet. The increasing demand of Bitcoin on the internet platform has made this paper to work out with the prediction of price rate of the Bitcoin day to day. Their study includes ML algorithm with the decision tree and regression algorithm. The comparison was made between the two algorithms, and the outcome of the regression algorithm is found to be more accurate than the decision tress.

Sovojit Manna et al. [8]. The paper works on the flight delay prediction by using the ML algorithm. Gradient booster decision tree is used for prediction of air traffic delay. Gradient booster decision tree can handle the regression model more effectively. The proposed feature can be used by the public, tourist, and airline agencies to obtain information about the delay in the flight in a very accurate manner.

AI Amin Biswas et al. [9]. Used a random forest regression method for detection of road accidents and number of causalities. The regression model first analyzes the dataset and then predicts the testing dataset or the real-time dataset. It is seen that the regression model provides a good performance.

Tessy Badriyah et al. [10]. The paper analyzes the stroke disease classification using ML. With the help of image processing techniques, the images are collected as the dataset. The Random forest algorithm uses the testing dataset and the real data for the prediction. Random forest algorithm proved to depict the highest accuracy.

K Vijyakumar et al. [11]. The proposed system works on the early prediction of diabetes in patients. The system is developed by the concept of ML algorithm with the Random Forest algorithm. The regression models are used for the classification and comparison of the dataset for the prediction. Hence, the algorithm depicts high accuracy in the results.

Sunakshi Mamgain et al. [12]. Author have proposed a model that works on the prediction of the popularity of cars in the market by considering the parameters such as price, maintenance, safety, mileage, etc. With the introduction of ML algorithm, Random Forest algorithm, and regression technique for prediction of the popularity of cars on the scaling 1–4.

Giorgio Biagetti et al. [13]. The paper works on the analysis of Bernstein polynomials for nonlinear system identification. Bernstein polynomial has a property that the co efficient is the value of the function to be approximated at point in fixed grid. Bernstein polynomial is particularly suitable to solve multivariate regression problems. The results show the accuracy of the technique for the identification of non-linear systems and provide a better performance with respect to standard techniques.

Swapanil Sayan Saha et al. [14]. Authors are concerned about the adulteration of fruits, vegetables, and fish with the hazards of formalin. The study works on the detection of formalin used at a high rate to keep the fruits, vegetables, and fish fresh for longer time, which causes health issues. The system model incorporates polynomial regression, linear regression, and Levenberg–Marquardt algorithm, which are supervised ML algorithm. The system provides an accurate prediction of the level of concentration of formalin at various levels of temperature. By the application of logistic regression and support vector machine, the system can identify artificially added and naturally formulated formalin.

Ching-Seh (Mike) Wu et al. [15]. The study focuses on the implementation of model with different ML techniques, such as regression and neural networks, which can predict the analysis of customers on the Friday sale: the customers that spent in the past and predicting the customers for the future with the accuracy to help the retail shop person plan how to control the crowd, manage his billings, and increase his sale.

Junheung Park et al. [16]. The study focuses on the psychological wellness indices with various parameters, such as anger, depression, stress, fatigue, etc. By predicting these indices, we can increase the practicality of psychological wellness and implement it in the medical system. The prediction model is developed by using four ML algorithms, namely, support vector regression, generalized regression neural networks, multilayer perception, and k nearest neighbor regression.

Different Types of Regression Algorithms in Machine Learning

The regression algorithms in ML are categorized in different types as given below:

1. Simple Linear Regression
2. Polynomial Regression
3. Support Vector Regression
4. Decision Tree Regression
5. Random Forest Regression

Simple Linear Regression

Linear regression is a measurable strategy that was presented by Sir Francis Galton in 1984 [6]. Linear regression is perhaps the most well-known and fascinating kind of regression strategy. Straightforward linear regression is a factual strategy that empowers to close and study connections between two distinctive quantitative factors. Linear regression is a direct model, a model that finds a direct connection between the information factors (A) and the single yield variable (B). Here, we foresee an objective variable (B) dependent on the information variable (A). A straight relationship should exist between target variable and indicator, thus comes the name linear regression. When there is a solitary info variable (A), the strategy is known as a straightforward linear regression. When there are different information factors, the technique is alluded as numerous linear regressions. Consider anticipating the compensation of a representative dependent on his/her age. We can without much of a stretch distinguish that there is by all accounts a relationship between representative's age and compensation (more the age, more is the compensation). Probably, the most mainstream utilizations of linear regression calculation are in monetary portfolio expectation, compensation gauging, land forecasts, and in rush hour gridlock in showing up at ETAs. The speculation of linear regression is

$$A = x + yB$$

A addresses pay, B is representative's age, and x and y are the coefficients of the condition. In this way, to anticipate A (compensation) given B (age), we need to know the estimations of x and y (the model's coefficients).

While preparing and assembling a regression model, these coefficients are learned and fitted to prepare information. The point of the preparation is to locate the best-fit line with the end goal that cost work is limited. The expense work helps in estimating the blunder. During the preparation cycle, we attempt to limit the mistake among real and anticipated qualities and in this manner, limiting the expense works. In Figure 5.1, the red focuses are the information and the blue line is the anticipated line for the preparation information. To get the anticipated worth, these information focuses are projected on to the line.

To sum up, our point is to discover such estimations of coefficients which will limit the expense work. The most widely recognized expense work is mean squared error (MSE), which is equivalent to the normal squared contrast between a perception's genuine and anticipated qualities. The coefficient esteems can be determined utilizing the Gradient Descent approach, which will be examined later in detail. To give a short agreement, in Gradient plunge, we start for certain arbitrary estimations of coefficients, process the inclination of

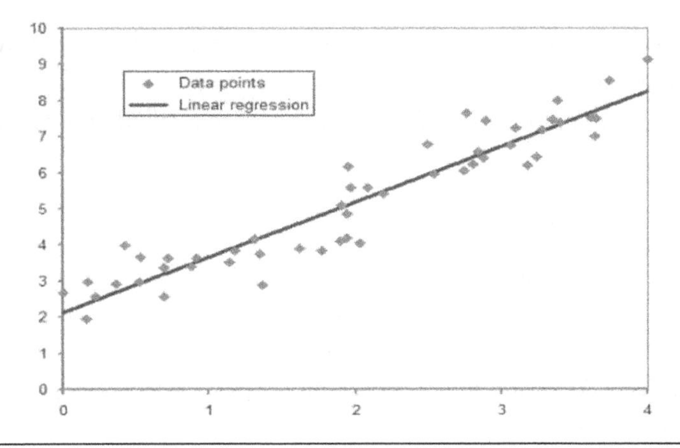

Figure 5.1 Linear Regression Graph

cost work on these qualities, update the coefficients, and compute the expense work once more. This cycle is rehashed until we locate a base estimation of cost work.

Significance of Linear Regression

The meaning of linear regression model is important for the accompanying underneath referenced reasons:

Descriptive – It helps in investigating the strength of the association between the outcome (dependent variable) and independent predictor variables.

Adjustment – It adapts with the impact of covariates or anomalies.

Predictors – It helps in foreseeing and assessing the future danger factors that influence the dependent variable.

Extent of prediction – It helps in investigating the degree of progress in the free information variable and the reliant yield variable.

Prediction – It helps in evaluating the new outcomes.

Simple Linear Regression Model in Health-Related Data Analysis In natural or health-related clinical information, linear regression model is frequently used to portray connections between two factors or among a few factors through measurable assessment. For instance, to know whether the probability of having high systolic BP (SBP) is affected by variables, for example, age, tallness, and weight, direct relapse is utilized. The variable to be clarified, i.e., SBP, is known as the needy variable, or on the other hand, the reaction factors that clarify it, age, tallness, weight, and gender are, called free factors.

Polynomial Regression

Non-linear information is generally experienced in a day-by-day life. Think about a portion of the conditions of movement as concentrated in material science and physics.

Projectile Motion: The stature of a projectile is calculated as $h = -½\, gt^2 + ut + ho$

Condition of movement under free fall: The distance went by an article in the wake of falling unreservedly under gravity for t seconds is ½ gt^2.

Distance traveled by a consistently accelerated body: The distance can be determined as

$$ut + 1/2at^2,$$

where,

g = acceleration due to gravity
u = beginning speed
ho = initial stature
a = acceleration

Notwithstanding these models, Non-linear patterns are additionally seen in the development pace of tissues, the advancement of infection pandemic, dark body radiation, the movement of the pendulum, and so forth. These models obviously demonstrate that we can't generally have a linear connection between the independent and dependent attributes. Henceforth, linear regression is a helpless decision for managing such nonlinear circumstances. This is the place where polynomial regression acts as the hero!!

Polynomial regression is an incredible method to experience the circumstances where a quadratic, cubic, or a more serious level nonlinear relationship exists. The basic idea in polynomial regression is to add forces of every autonomous independent variable as new attributes and afterward train a linear model on this extended assortment of highlights.

In polynomial regression, we change the first highlights into polynomial highlights of a given degree and afterward apply linear

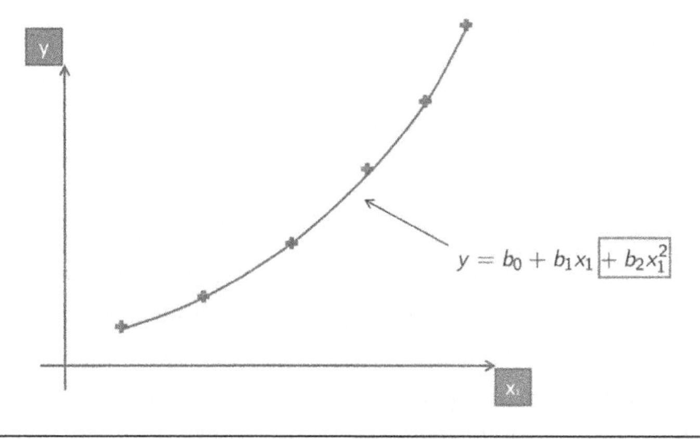

Figure 5.2 Polynomial Regression Graph

regression on it. Think about the above straight model: $Y = a + bX$ is changed into something like

$$Y = a + b\ X + c\ X^2$$

It is as yet a linear model; however, the curve is currently quadratic as opposed to a line. scikit-learn gives polynomial features class to change the highlights. Consequently, if we attempt to utilize a straightforward linear regression in the given chart, then the linear regression line will not fit well overall. It is extremely hard to fit a linear regression line in the diagram given in Figure 5.2 with a low estimation of mistake. Thus, we can attempt to utilize the polynomial regression to fit a polynomial line so that we can accomplish a base blunder or least expense work (Figure 5.2).

On the off chance that we increment the degree to a high worth, the curve becomes overfitted as it learns the noise and outliers in the information also.

Advantages of Using Polynomial Regression:

- Polynomial gives the best estimation of the connection between the dependent and independent factors.
- A broad range of functions can be fit under it.
- Polynomial fundamentally fits a wide scope of the curve.

Disadvantages of Using Polynomial Regression:

- The presence of a couple of anomalies in the information can truly influence the aftereffects of the nonlinear examination.
- These are excessively touchy to the exceptions.

- In expansion, there are lamentably fewer model approval apparatuses for the recognition of anomalies in nonlinear regression than there are for linear regression.

Polynomial Regression Model in COVID-19 Growth Pattern Analysis
Coronavirus is spreading inside such a gigantic scourge for the globe. This plague taints a ton of people in Egypt. The world wellbeing association states that COVID-19 could be spread starting with one individual then onto the next at an exceptionally quick speed through contact and respiratory splash. Nowadays, Egypt and all nations worldwide ascend to a compelling advance to examine this illness and kill the impacts of this pestilence. In this paper, the genuine information base of Coronavirus for Egypt has been broken down from February 15, 2020 to June 15, 2020, and anticipated with the quantity of patients that will be tainted with Coronavirus, and assessed the scourge last size. A few relapse examinations models have been applied for information investigation of Coronavirus in Egypt. In this investigation, we've applied seven relapse examination-based models that are outstanding polynomial, quadratic, third-degree, fourth-degree, fifth-degree, sixth-degree, and logit development individually for the COVID-19 dataset. Consequently, the remarkable fourth-degree, fifth-degree, and sixth-degree polynomial relapse models are astounding, unlike fourth-degree model that will help the public authority in setting up its methodology for one month. What's more! We have applied the notable logit development relapse model and acquired the accompanying epidemiological bits of knowledge: right off the bat, the plague pinnacle might actually reach on June 22, 2020 and final season of pestilence on September 8, 2020. Besides, the last all-out size for cases 1.6676eþ05 cases. The activity of legislature of inter-evention over a generally long stretch is important to limit the last plague size.

Support Vector Machine Regression

In ML and data analytics, support vector machines (SVMs, also support-vector organizations) are supervised learning models with associated learning algorithms that analyze and investigate pattern regression analysis based on classifications. It was proposed at AT&T Bell Laboratories by Vapnik et al. In Support Vector Machine Regression (SVR), we distinguish a hyperplane with most extreme edge to such an extent that the greatest number of information focus is inside that edge. SVRs are practically like the SVM arrangement calculation. SVM is another most remarkable calculation with solid hypothetical establishments dependent on the Vapnik–Chervonenkis hypothesis, as characterized by oracle docs. This supervised ML algorithm has solid regularization and can be utilized both for classification or regression challenges. They are characterized by usage of kernels, the sparseness of the solution, and the capacity control gained by acting on the margin, or on number of support vectors, etc. The capacity of the system is controlled by parameters that do not depend on the dimensionality of feature space. Since the SVM algorithm operates natively on numeric attributes, it uses a z-score normalization on numeric attributes. In regression, SVM algorithms use margin of loss function to solve regression problems.

Rather than limiting the mistake rate as in straightforward linear regression, we attempt to fit the blunder inside a specific edge. Our goal in SVR is to fundamentally consider the focuses that are inside the edge. Our best fit line is the hyperplane that has the most extreme number of focuses. In SVM algorithm, "c" margin is always considered as a variation margin.

In Figure 5.3, red (top and bottom) lines represent the patterns with the margin "c" in the positive and negative axes margin.

SVM regression algorithms calculations have discovered a few applications in the oil and gas industry, order of pictures, and text and hypertext arrangement. In the oil fields, it is explicitly utilized for

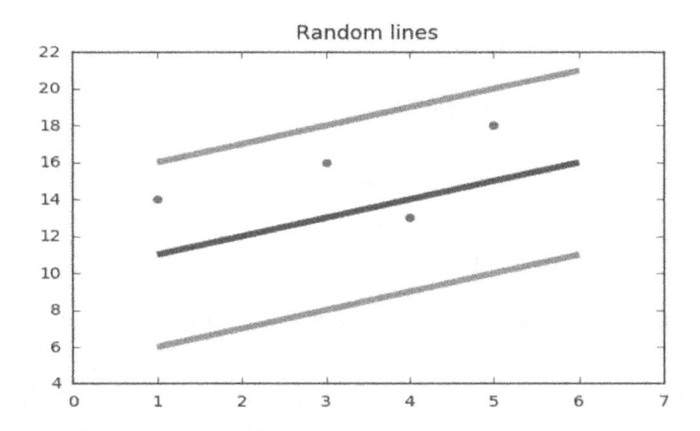

Figure 5.3 Support Vector Regression with Margin

investigation to comprehend the situation of layers of rocks and make 2-D and 3-D models as a portrayal of the earth.

Decision Tree Regression

Decision trees can be utilized for classification and characterization just as regression. In Decision trees, at each level, we need to distinguish the division property. On account of regression, the ID3 calculation can be utilized to distinguish the parting hub by lessening the standard deviation (in order data classification is utilized).

A decision tree is worked by apportioning the information into subsets containing examples with comparable qualities (homogenous). Standard deviation is utilized to compute the homogeneity of a mathematical example. On the off chance that the mathematical example is totally homogeneous, its standard deviation is zero.

The steps for finding the splitting node are briefly described below:

1. Calculate the standard deviation of the target variable using the below formula.

$$\sigma = \sqrt{\frac{1}{N} \sum_{i=1}^{N} (x_i - \mu)^2}$$

2. Split the dataset on various attributes and figure the standard deviation for each branch (standard deviation for target and indicator). This worth is deducted from the standard deviation before the split. The outcome is the standard deviation decrease.

$$\text{SDR}(T, X) = S(T) - S(T, X)$$

3. The attribute with the biggest standard deviation decrease is picked as the partition node.
4. The dataset is partitioned depending on the estimations of the chose trait. This interaction is run recursively on the non-leaf branches until all information is handled.

To maintain a strategic distance from overfitting, the coefficient of deviation (CV) is utilized, which chooses when to quit spreading. At last, the normal of each branch is allotted to the connected leaf hub (in relapse mean is taken though in grouping method of leaf nodes is taken).

Random Forest Regression

Arbitrary backwoods are a gathering approach where we consider the forecasts of a few choice regression trees.

1. Select An arbitrary focuses.
2. Identify *n* where *n* is the quantity of choice tree regressors to be made. Repeat Stages 1 and 2 to make a few relapse trees.
3. The normal of each branch is appointed to the leaf node in every decision tree.
4. To foresee yield for a variable, the normal of the multitude of forecasts of all choice trees is mulled over.

Arbitrary Forest forestalls overfit (which is regular in choice trees) by making irregular subsets of the highlights and building more modest trees utilizing these subsets.

Case Study on Air Quality Prediction Using Machine Learning Regression Techniques

Air quality checking and forecast have got perhaps the most fundamental movement in numerous mechanical and metropolitan territories. The nature of air is exceptionally influenced because of the different types of poisons. With expanding air contamination, productive air quality observing models is to be actualized; these models gather data about the centralization of air toxins. We propose a general and viable way to deal with care of the three issues: expectation, insertion, and highlight examination. These three issues were settled already utilizing three unique models; however, now in the proposed framework, we can tackle these three issues in a single model called the Deep Air Learning (DAL). This methodology is related to unlabeled spatio-fleeting information to improve the presentation of insertion and the expectation of air quality. We assess our methodology by the examinations dependent on ongoing information sources acquired by the Karnataka State Pollution Control Board (KSPCB), India. The point of this exploration is to examine different computerized reasoning and ML-based procedures for checking and anticipating

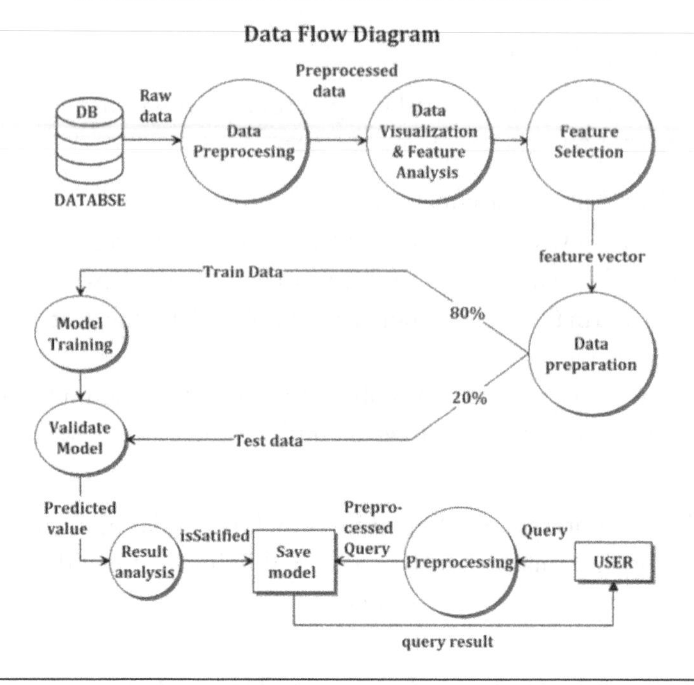

Figure 5.4 Air Quality Prediction Model

the air quality. Python language is used for the development of the experimentation.

Figure 5.4 represents the air quality prediction system in a detailed manner by representing how the data from database are captured, validated, and processed by the prediction system till it reaches the final step, i.e., generating the query result.

Figure 5.5 shows the representation of how python code imports libraries. Figure 5.6 represents certain columns of parameter values for different index variables.

Figure 5.7 represents the AQI range with various AQI categories defined that are considered during the experimentation work.

Figure 5.8 represents data labeling. In Figure 5.9, data graphs are plotted after the data analytics.

Figure 5.10 represents the data split for testing and training datasets.

Figure 5.11 shows the air quality analysis, its parameters, and its range taken for the analysis.

In Figure 5.12, AQI category, pollutants, breakpoints, and health impacts are represented after the data analysis.

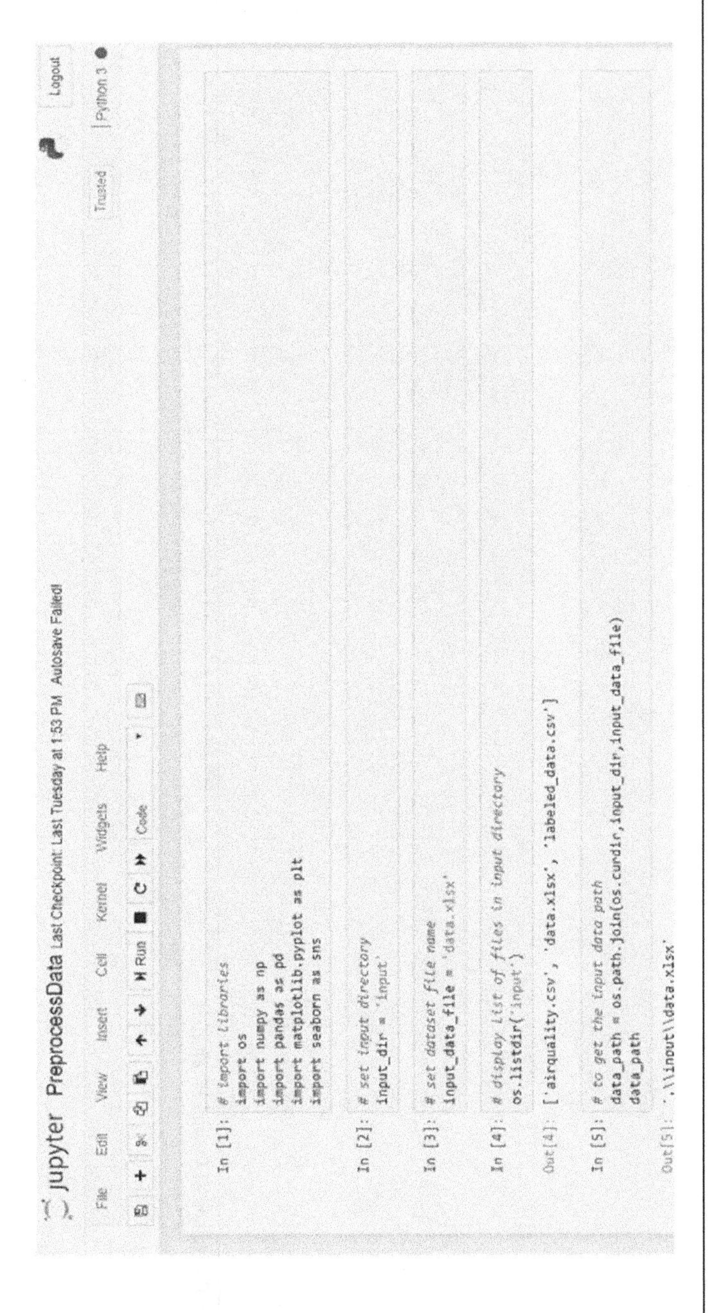

Figure 5.5 How Python Code Imports Libraries

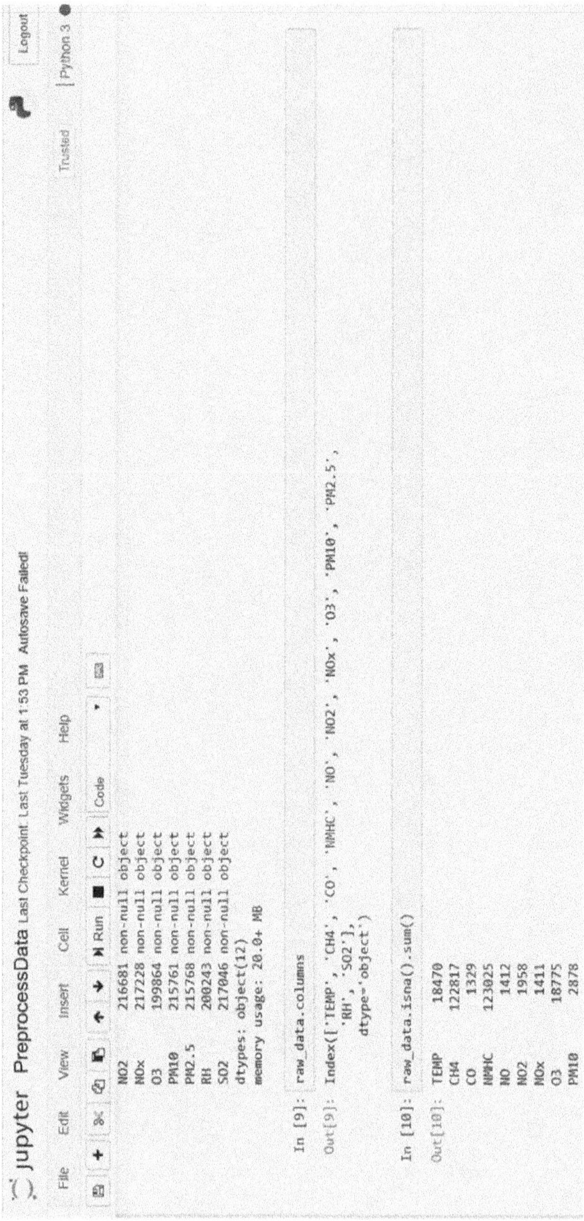

```
jupyter PreprocessData Last Checkpoint: Last Tuesday at 1:53 PM Autosave Failed!
File Edit View Insert Cell Kernel Widgets Help                Trusted | Python 3

NO2       216681 non-null object
NOx       217228 non-null object
O3        199864 non-null object
PM10      215761 non-null object
PM2.5     215768 non-null object
RH        200243 non-null object
SO2       217046 non-null object
dtypes: object(12)
memory usage: 20.0+ MB

In [9]: raw_data.columns

Out[9]: Index(['TEMP', 'CH4', 'CO', 'NMHC', 'NO', 'NO2', 'NOx', 'O3', 'PM10', 'PM2.5',
               'RH', 'SO2'],
              dtype='object')

In [10]: raw_data.isna().sum()

Out[10]: TEMP      18470
         CH4      122817
         CO         1329
         NMHC     123025
         NO         1412
         NO2        1958
         NOx        1411
         O3        18775
         PM10       2878
```

Figure 5.6 Columns of Certain Parameters

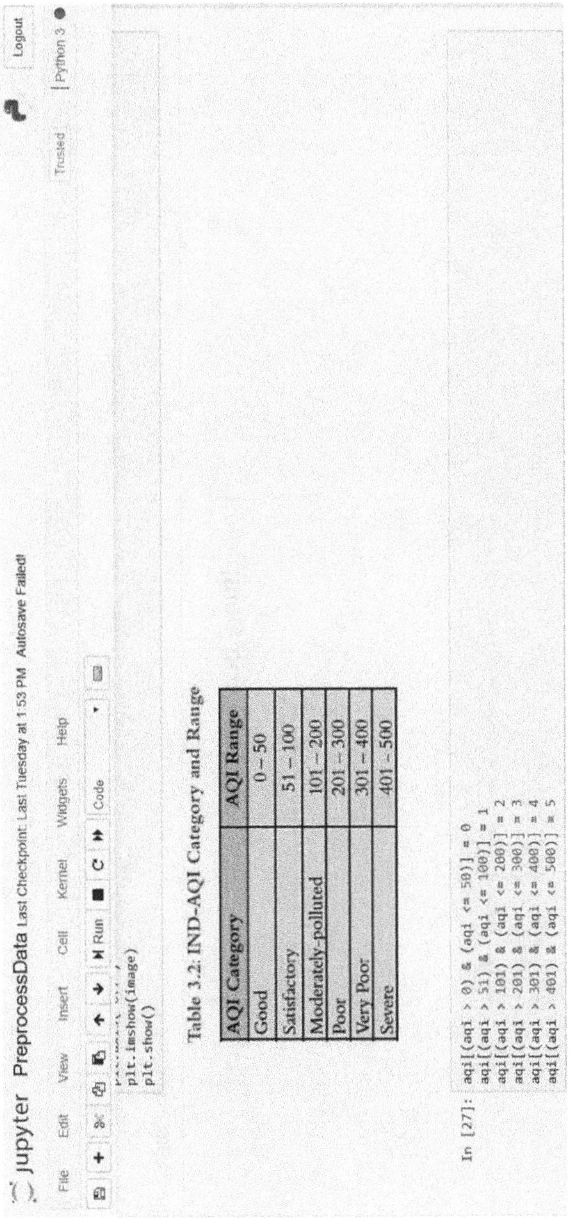

jupyter PreprocessData Last Checkpoint: Last Tuesday at 1:53 PM Autosave Failed!

File Edit View Insert Cell Kernel Widgets Help

Trusted | Python 3

Logout

```
plt.imshow(image)
plt.show()
```

Table 3.2: IND-AQI Category and Range

AQI Category	AQI Range
Good	0 – 50
Satisfactory	51 – 100
Moderately-polluted	101 – 200
Poor	201 – 300
Very Poor	301 – 400
Severe	401 – 500

```
In [27]: aqi[(aqi > 0) & (aqi <= 50)] = 0
         aqi[(aqi > 51) & (aqi <= 100)] = 1
         aqi[(aqi > 101) & (aqi <= 200)] = 2
         aqi[(aqi > 201) & (aqi <= 300)] = 3
         aqi[(aqi > 301) & (aqi <= 400)] = 4
         aqi[(aqi > 401) & (aqi <= 500)] = 5
```

Figure 5.7 Air Quality Index Category and Range

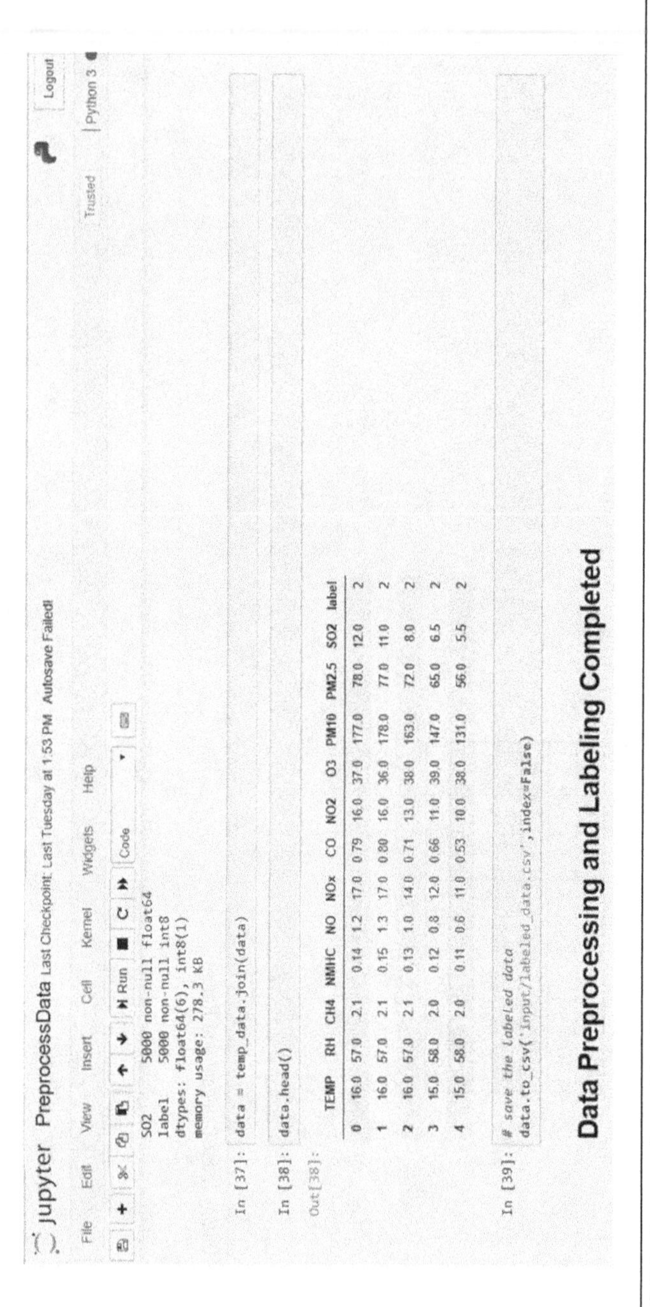

Figure 5.8 Data Preprocessing and Labeling

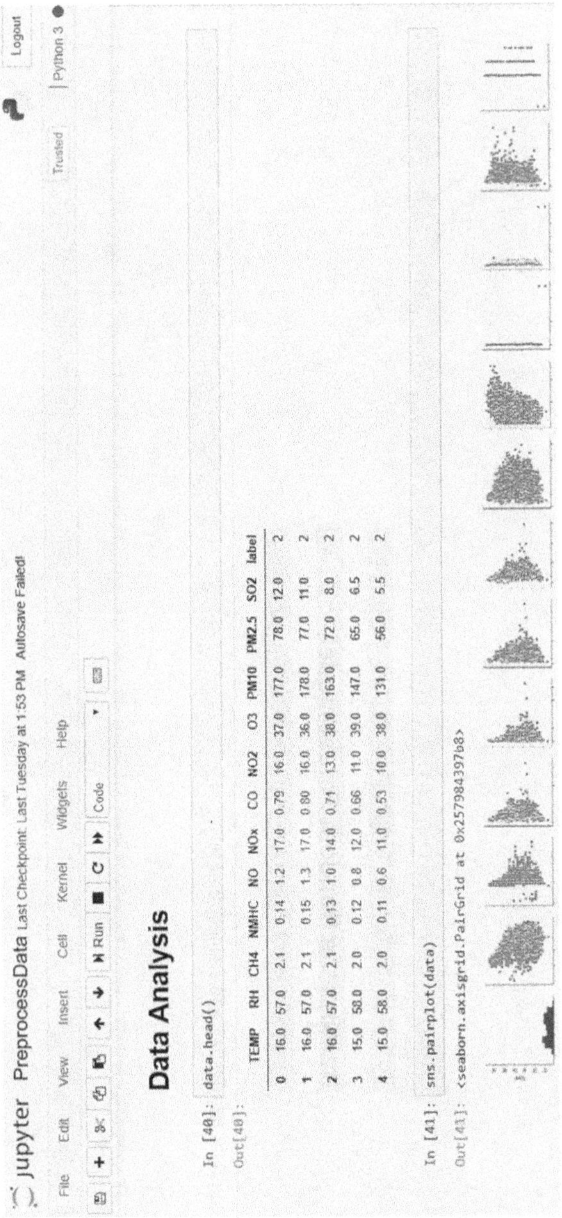

jupyter PreprocessData Last Checkpoint: Last Tuesday at 1:53 PM Autosave Failed!

Logout

File Edit View Insert Cell Kernel Widgets Help Trusted Python 3

Run Code

Data Analysis

In [40]:
```
data.head()
```

Out[40]:

	TEMP	RH	CH4	NMHC	NO	NOx	CO	NO2	O3	PM10	PM2.5	SO2	label
0	16.0	57.0	2.1	0.14	1.2	17.0	0.79	16.0	37.0	177.0	78.0	12.0	2
1	16.0	57.0	2.1	0.15	1.3	17.0	0.80	16.0	36.0	178.0	77.0	11.0	2
2	16.0	57.0	2.1	0.13	1.0	14.0	0.71	13.0	38.0	163.0	72.0	8.0	2
3	15.0	58.0	2.0	0.12	0.8	12.0	0.66	11.0	39.0	147.0	65.0	6.5	2
4	15.0	58.0	2.0	0.11	0.6	11.0	0.53	10.0	38.0	131.0	56.0	5.5	2

In [41]:
```
sns.pairplot(data)
```

Out[41]: <seaborn.axisgrid.PairGrid at 0x257984397b8>

Figure 5.9 Data Analysis Stage and Data Graph

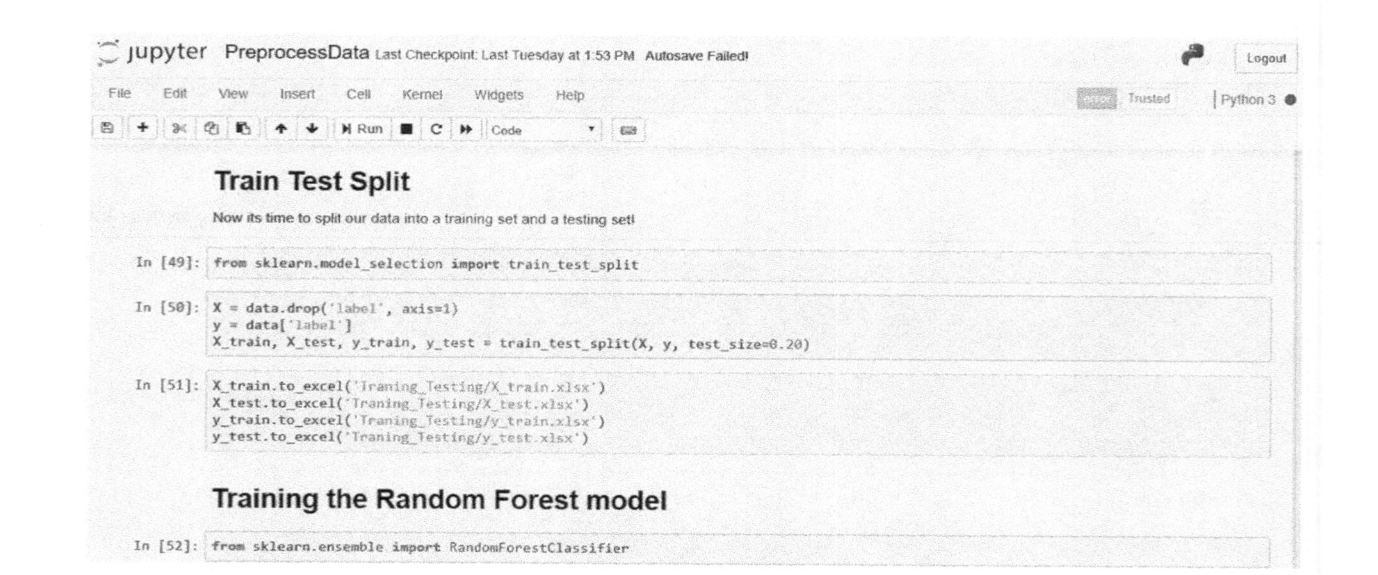

Figure 5.10 Data Split into Testing and Training Dataset

jupyter PreprocessData Last Checkpoint: Last Tuesday at 1:53 PM (autosaved)

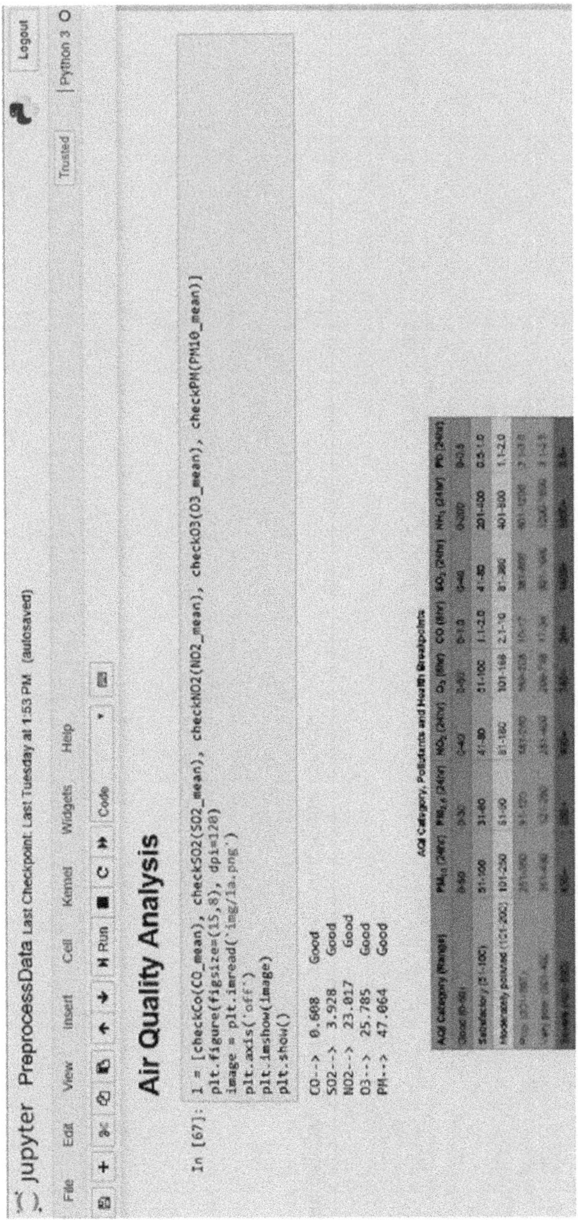

File Edit View Insert Cell Kernel Widgets Help Trusted Python 3

Air Quality Analysis

```
In [67]: l = [checkCo(CO_mean), checkSO2(SO2_mean), checkNO2(NO2_mean), checkO3(O3_mean), checkPM(PM10_mean)]
         plt.figure(figsize=(15,8), dpi=120)
         image = plt.imread('img/1a.png')
         plt.axis('off')
         plt.imshow(image)
         plt.show()

CO--> 0.608    Good
SO2--> 3.928    Good
NO2--> 23.017    Good
O3--> 25.785    Good
PM--> 47.064    Good
```

AQI Category, Pollutants and Health Breakpoints

AQI Category (Range)	PM_{10} (24hr)	$PM_{2.5}$ (24hr)	NO_2 (24hr)	O_3 (8hr)	CO (8hr)	SO_2 (24hr)	NH_3 (24hr)	Pb (24hr)
Good (0-50)	0-50	0-30	0-40	0-50	0-1.0	0-40	0-200	0-0.5
Satisfactory (51-100)	51-100	31-60	41-80	51-100	1.1-2.0	41-80	201-400	0.5-1.0
Moderately polluted (101-200)	101-250	61-90	81-180	101-168	2.1-10	81-380	401-800	1.1-2.0
[illegible]	[illegible]	[illegible]	[illegible]	[illegible]	[illegible]	[illegible]	[illegible]	[illegible]
[illegible]	[illegible]	[illegible]	[illegible]	[illegible]	[illegible]	[illegible]	[illegible]	[illegible]
[illegible]	[illegible]	[illegible]	[illegible]	[illegible]	[illegible]	[illegible]	[illegible]	[illegible]

Figure 5.11 Air Quality Analysis, Parameters, and Its Range

AQI Category, Pollutants and Health Breakpoints

AQI Category (Range)	PM_{10} (24hr)	$PM_{2.5}$ (24hr)	NO_2 (24hr)	O_3 (8hr)	CO (8hr)	SO_2 (24hr)	NH_3 (24hr)	Pb (24hr)
Good (0-50)	0-50	0-30	0-40	0-50	0-1.0	0-40	0-200	0-0.5
Satisfactory (51-100)	51-100	31-60	41-80	51-100	1.1-2.0	41-80	201-400	0.5-1.0
Moderately polluted (101-200)	101-250	61-90	81-180	101-168	2.1-10	81-380	401-800	1.1-2.0
Poor (201-300)	251-350	91-120	181-280	169-208	10-17	381-800	801-1200	2.1-3.0
Very poor (301-400)	351-430	121-250	281-400	209-748	17-34	801-1600	1200-1800	3.1-3.5
Severe (401-500)	430+	250+	400+	748+	34+	1600+	1800+	3.5+

AQI	Associated Health Impacts
Good (0-50)	Minimal impact
Satisfactory (51-100)	May cause minor breathing discomfort to sensitive people.
Moderately polluted (101–200)	May cause breathing discomfort to people with lung disease such as asthma, and discomfort to people with heart disease, children and older adults.
Poor (201-300)	May cause breathing discomfort to people on prolonged exposure, and discomfort to people with heart disease.
Very poor (301-400)	May cause respiratory illness to the people on prolonged exposure. Effect may be more pronounced in people with lung and heart diseases.
Severe (401-500)	May cause respiratory impact even on healthy people, and serious health impacts on people with lung/heart disease. The health impacts may be experienced even during light physical activity.

Figure 5.12 AQI Category, Pollutants, Breakpoints, and Health Impacts

Conclusions and Future Work

Machine learning (ML) is a very incredible and evitable tool. ML algorithms are very useful for data analysis and future information forecast. In this work, various ML regression algorithms are discussed in detail. The algorithms are discussed with various available real-time problem scenarios. This work considers only the supervised ML-based methods. The case study elaborates the regression algorithm for the air pollution control analysis with various samples of real datasets. The proposed work is useful for future air quality and pollution control based on data analysis techniques and shows the effectiveness for the future research domain.

References

[1] Pahwa, K., and Agarwal, N., "Stock Market Analysis Using Supervised Machine Learning," *2019 International Conference on Machine Learning, Big Data, Cloud and Parallel Computing (COMITCon)*, Faridabad, India, 2019, pp. 197–200, doi: 10.1109/COMITCon.2019.8862225.

[2] Yang, Zhixia, Deng, Naiyang, and Tian Yingjie, "A Multi-class Classification Algorithm Based on Ordinal Regression Machine," *International Conference on Computational Intelligence for Modelling, Control and Automation and International Conference on Intelligent Agents, Web Technologies and Internet Commerce (CIMCA-IAWTIC'06)*, Vienna, 2005, pp. 810–815, doi: 10.1109/CIMCA.2005.1631568.

[3] Gambhir, E., Jain, R., Gupta, A., and Tomer, U., "Regression Analysis of COVID-19 Using Machine Learning Algorithms," *2020 International Conference on Smart Electronics and Communication (ICOSEC)*, Trichy, India, 2020, pp. 65–71, doi: 10.1109/ICOSEC49089.2020.9215356.

[4] Rong, Shen, and Bao-wen, Zhang. "The Research of Regression Model in Machine Learning Field," *MATEC Web of Conferences*, Vol. 176, No. 01033 (2018), https://doi.org/10.1051/matecconf/201817601033, IFID 2018.

[5] Lee, Ki-Young, Kim, Kyu-Ho, Kang, Jeong-Jin, Choi, Sung-Jai, Im, Yong-Soon, Lee, Young-Dae, and Lim, Yun-Sik, "Comparison and Analysis of Linear Regression & Artificial Neural Network." *International Journal of Applied Engineering Research*, ISSN 0973-4562 Vol. 12, No. 20 (2017), pp. 9820–9825.

[6] Kumari, Khushbu, and Yadav, Suniti, "Linear Regression Analysis Study," *Journal of the Practice of Cardiovascular Sciences*, Vol. 4, No. 1 (January–April 2018), pp. 33–36.

[7] Rathan, K., Sai, S. V., and Manikanta, T. S. (2019). Crypto-Currency Price Prediction Using Decision Tree and Regression Techniques. *2019 3rd International Conference on Trends in Electronics and Informatics (ICOEI)*. doi: 10.1109/icoei.2019.8862585.
[8] Manna, S., Biswas, S., Kundu, R., Rakshit, S., Gupta, P., and Barman, S. (2017). "A Statistical Approach to Predict Flight Delay Using Gradient Boosted Decision Tree." *2017 International Conference on Computational Intelligence in Data Science(ICCIDS)*. doi: 10.1109/iccids.2017.8272656
[9] Biswas, A. A., Mia, M. J., and Majumder, A., "Forecasting the Number of Road Accidents and Casualties Using Random Forest Regression in the Context of Bangladesh." *2019 10th International Conference on Computing, Communication and Networking Technologies (ICCCNT)*, 2019, doi: 10.1109/icccnt45670.2019.8944500
[10] Badriyah, T., Sakinah, N., Syarif, I., and Syarif, D. R. Machine Learning Algorithm for Stroke Disease Classification. 2020 International Conference on Electrical, Communication, and Computer Engineering (ICECCE), 2020, doi:10.1109/icecce49384.2020.9179307
[11] VijiyaKumar, K., Lavanya, B., Nirmala, I., and Caroline, S. S. (2019). Random Forest Algorithm for the Prediction of Diabetes. 2019 IEEE International Conference on System, Computation, Automation and Networking (ICSCAN). doi:10.1109/icscan.2019.8878802
[12] Mamgain, S., Kumar, S., Nayak, K. M., and Vipsita, S. (2018). Car Popularity Prediction: A Machine Learning Approach. 2018 Fourth International Conference on Computing Communication Control and Automation (ICCUBEA). doi:10.1109/iccubea.2018.8697832
[13] Biagetti, G., Crippa, P., Falaschetti, L., and Turchetti, C. (2017). Machine learning regression based on particle bernstein polynomials for nonlinear system identification. 2017 IEEE 27th International Workshop on Machine Learning for Signal Processing (MLSP). doi:10.1109/mlsp.2017.8168148
[14] Saha, S. S., Siraj, M. S., and Habib, W. B. (2017). FoodAlytics: A Formalin Detection System Incorporating a Supervised Learning Approach. *2017 IEEE Region 10 Humanitarian Technology Conference (R10-HTC)*. doi: 10.1109/r10-htc.2017.8288898
[15] Wu, C. M., Patil, P., and Gunaseelan, S., "Comparison of Different Machine Learning Algorithms for Multiple Regression on Black Friday Sales Data," *2018 IEEE 9th International Conference on Software Engineering and Service Science (ICSESS)*, Beijing, China, 2018, pp. 16–20, doi: 10.1109/ICSESS.2018.8663760.
[16] Park, J., Kim, K., and Kwon, O., "Comparison of Machine Learning Algorithms to Predict Psychological Wellness Indices for Ubiquitous Healthcare System Design," *Proceedings of the 2014 International Conference on Innovative Design and Manufacturing (ICIDM)*, Montreal, QC, 2014, pp. 263–269, doi: 10.1109/IDAM.2014.6912705.

6

Machine Learning-Based Industrial Internet of Things (IIoT) and Its Applications

M. SURESH KUMAR[1], S. RACHEL[1], AND M. V. KAVISELVAN[2]

[1]*Department of Information Technology, Sri Sairam Engineering College, Chennai, India*
[2]*Department of Mechanical Engineering, Sri Sairam Engineering College, Chennai, India*

Contents

DOI: 10.1201/9781003119838-6

History of Internet of Things

The Internet of Things [1] was started in the early 1980s, but it was not publicly called until 1999. The first example of an IoT is a Coca Cola machine, based at Carnegie Mellon University. Local programmers can link it to the refrigerator appliance through the internet and check to see if a drink was available and if it was cold before the ride.

By 2013, IoT had developed into a device using different technologies, ranging from wireless networking, embedded devices, networks of wireless sensors, GPS, control systems, etc.

Kevin Ashton, Executive Director and Visiting Engineer, is known as the Father of IoT, at the Massachusetts Institute of Technology (MIT). He led the work on coding for the next decade. The Internet of Things word was invented by Jeremy Cowan. The first IoT system connected to a computer via TCP/IP protocol was toaster.it.

Figure 6.1 Applications of IoT

IoT works in nearly every industry, namely, financial services, mobile banking, health care, etc. IoT shocked us with so many things we had never imagined before. Everything changed, from automatic home security to smart transportation. Here, we will identify the sectors that are booming with IoT.

It is a matter of fact today that a vast number of users have surpassed the bulk of IoT products on this planet. There are actually 7.62 billion people on our planet, but to your horror, by 2021, with a growing IoT system graph, nearly 20 billion IoT smart devices will be up and running with a surge in demand for 5G networks (Figure 6.1).

Scope of IoT

The rapid growth of technology such as embedded systems, wireless sensor technology, machine learning (ML), and automation plays vital role in IoT development. This emerging technology has a greater scope in many technological fields, including the following:

- Automotive and transportation industry
- Telecommunication
- Health care
- Logistics
- Supply chain
- Agriculture
- Wearable devices

- Smart cites and smart home
- Smart retail.

Features of IOT

Following are the seven crucial characteristics of IOT:

- Connectivity
- Things
- Data
- Communication
- Intelligence
- Action
- Ecosystem

Leading Types of IoT

- LPWAN (Low-Power Wide AreaNetwork)
- Cellular (3G/4G/5G)
- Zigbee and other meshprotocols
- Bluetooth and BLE
- Wi-Fi
- RFID (Radio Frequency Identification).

Advantages of IoT

IoT is going to bring many revolutionary changes in this new technological era, and it has many advantages, such as great number of new business opportunities, new capabilities to predict and act, improved monitoring, fine-tuned services and products, improved control of operation process, and the most essential, starting of new revenue. There are also many other advantages to this technology.

Eventhough IoT has many advantages, it has few disadvantages also. The main three disadvantages are:

- Breach of privacy
- Over-reliance on technology
- Loss of jobs

Machine Learning and Deep Learning

Machine Learning (ML) is a subset of artificial intelligence (AI) technology. It is simply defined as the study of computer algorithms that can be updated automatically by self-learning process. ML improves the dataset by self-learning process. ML is used to collect the various types of data while working and all the data are analyzed and organized by a cloud storage system. From these gathered real-time data, ML improves its efficiency and provides accurate actions depending on the data analyzed. The learning process of this system is majorly divided into three types:

- Supervised learning
- Unsupervised learning
- Semi-supervised learning

All these processes improve the idea of the system with less human intervention. In supervised learning, the example input and the goal of operation are provided to the system. Thus it will be easy for the system to collect the real-time data from the process of operation and improve the idea of the system.

In unsupervised learning, the example input and the goal of the process will not be provided to the system. In this type of learning, ML will improve its process only by gathering the real-time information from the operation. This type of learning is quite difficult to analyze the data collected and execute the preferred action on time.

Semi-supervised learning is the moderate learning process of ML technology. In this learning process, some of the example inputs and goal of operation will be provided to the system; thus, it obtains the other needed data from the time of operation. This is easier than unsupervised learning and harder than supervised learning process (Figure 6.2).

Applications of ML

- Image recognition
- Speech recognition
- Traffic prediction
- Product recommendations

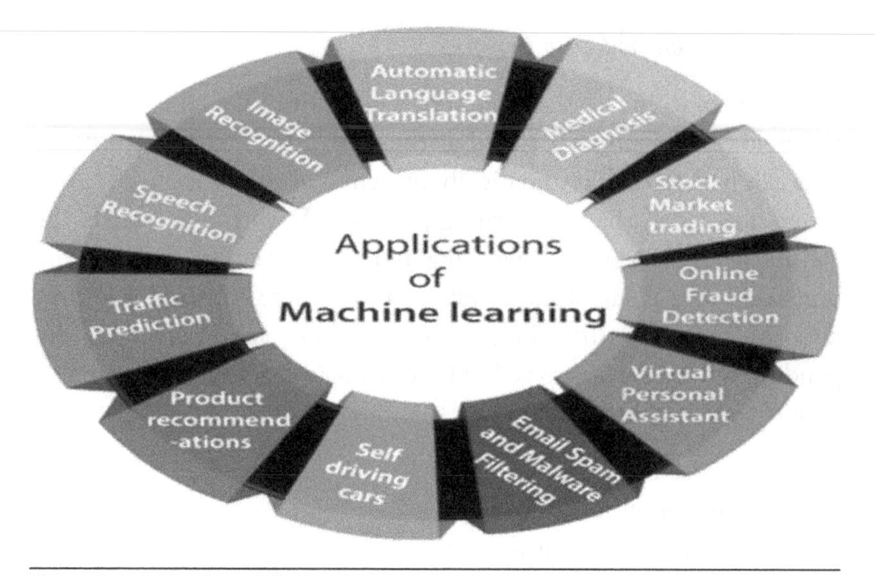

Figure 6.2 Applications of Machine Learning

- Self-driving cars
- Email spam and malware filtering
- Virtual personal assistant
- Online fraud detection

Except the above-mentioned applications, ML also has wide range of applications in our day-to-day life. Nowadays, ML is one of the most awaited technologies for changing the industrial world and it will also create a great impact on the Industry 4.0. ML has the ability of correcting its own problems by self-learning and finding an appropriate solution on its own, thus providing all possible solutions for many problems without the intervention of human being (Figure 6.2).

IIoT – Industrial Internet of Things

The Industrial Internet of Things (IIoT) relates to the web of integrated devices, such as sensors and robots, and others, like production and energy storage, through a network of industrial applications for computers. This interconnectivity facilitates data storage and sharing services. The transmitted data are then processed, potentially

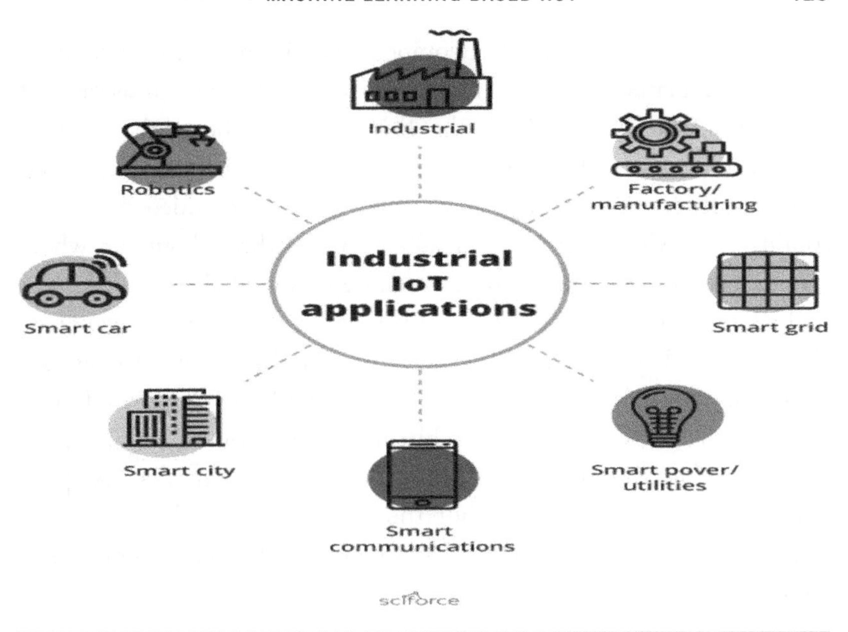

Figure 6.3 Industrial IoT Applications

promoting changes in the industry's competitiveness and performance, as well as other economic benefits.

Industrial organizations are increasingly investing in IIoT to enhance operational visibility and insights, which will help streamline manufacturing processes. Taking the complexity out of connecting, deploying and managing devices within the field are vital to IoT success (Figure 6.3).

Industrial modernization and therefore the shift to smart manufacturing have sparked innovations in automation and robotics and therefore the Industrial Internet of Things (IIoT). Introducing artificial intelligence (AI), interoperability and secure connectivity enable real-time monitoring, control and optimization of processes, resources and systems – all of which result in greater productivity, increased safety, and reduced cost.

History of IIoT

The history of the IIoT began with Dick Morley's invention of the programmable logic controller (PLC) in 1968, which was used by General Motors in their automatic transmission manufacturing division.

These PLCs allowed individual components within the manufacturing chain to be better controlled. Honeywell and Yokogawa implemented the world's first DCSs in 1975, the TDC 2000 and hence the Centum system, respectively. These DCSs were subsequent steps in facilitating scalable process management in a facility, with the added benefit of contingency redundancies by spreading control throughout the whole system, preventing a single point of failure in a central space.

The current conception of the IIoT arose after the emergence of cloud technology in 2002, which allows information storage for historical trends, and so the event of the Open Platform Communication (OPC) Unified Architecture protocol in 2006, which enabled secure and remote communications between devices, programs, and data sources without the requirement of human intervention or interface. The OPC server is a software program that converts the hardware communication protocol employed by a PLC into the OPC protocol.

Advantages of IIoT

The phrase Industrial Internet of Things, pertaining to the economic branch of the IoT, is commonly located within the manufacturing industries. Improved efficiency, analytics, and the development of the workplace are future advantages of the economic IoT. The potential for growth by introducing IIoT is expected to reach $15 trillion of global GDP by 2030. Through IoT apps comes the possibility of automation. Smart offices use a number of connected devices to track, control, and handle multiple activities within an organization. They may also assist staff in managing their schedules in order to be more efficient, using their time efficiently and successively.

Enabling a $800 billion global market by 2024, IIoT is going to be a serious catalyst for connected products, services, and solutions.

IIoT in Industry 4.0

Indudtry 4.0 is a phrase coined in Europe. It means as equivalent as IIoT and refers to the fourth technological revolution. The term is interchangeable with IIoT and is now recognized globally.

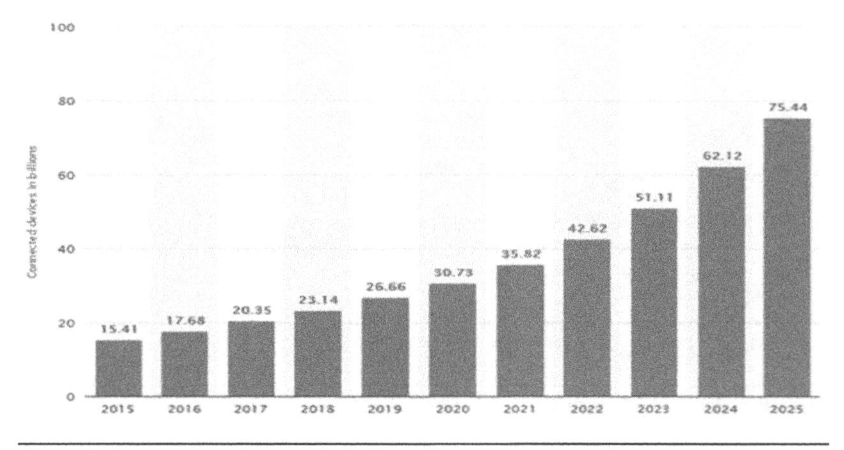

Figure 6.4 Number of IoT devices in 2015–2025

In order to form a more comprehensive and better integrated environment for businesses specializing in manufacturing and providing chain management, IIoT or smart manufacturing blends physical development and practices with smart digital technologies, ML, and enormous data (Figure 6.4).

Sensor data, computer communications, and automation systems are referred to in the IIoT. It is more descriptive than the general phrase Business 4.0, which covers the industry's entire digital revolution.

IIoT use cases are the latest features known in Business 4.0, supporting the smart industry. In the economic focus of Industry 4.0, these capabilities are basically what we often call usage cases for IoT or IIoT:

- track and trace
- structural health monitoring
- remote diagnosis
- control

They're all typical use cases.

IoT and IIoT

In terms of availability, intelligent and wired computers have an identical main character. Its general uses are the only distinction

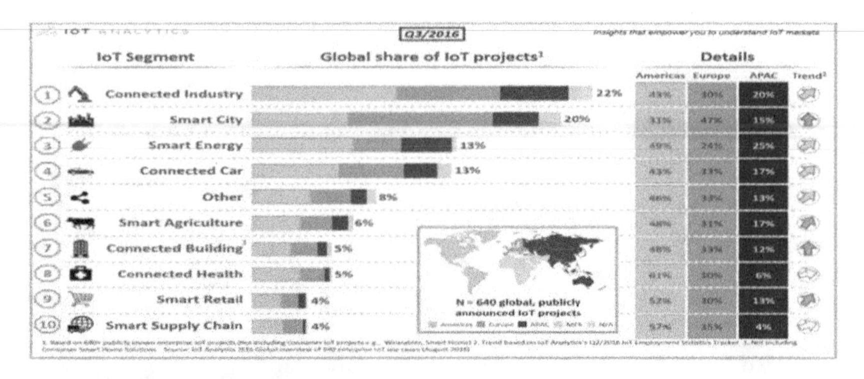

Figure 6.5 Market of IoT devices

between the two. Though IoT is most widely used for consumer use, IIoT is used for distribution, supply chain control, and management applications for industrial purposes.

List of the simplest IoT companies is given below:

- Vates
- ScienceSoft (USA and Europe)
- Oxagile (New York, USA)
- Style Lab IoT Software Company (San Francisco, CA)
- HQ_Software Industrial IoT Company (USA and Europe)
- PTC (Boston, MA)
- Cisco (San Jose, CA)
- ARM IoT Security Company (Cambridge, Cambs) (Figure 6.5).

The Future of IIoT

As the IIoT progresses, it creates unparalleled opportunities for companies to simplify processes, maximize consumer service, and accelerate substantial sales gains from the SMB to the global business. In order to tap into the business advantages of the linked facility, the long-term IIoT would focus more on predictive maintenance, better system connectivity, and cheaper access for enterprises of all sizes.

IIoT Use Cases

The IIoT is the network of a multitude of industrial devices linked by communication technology, culminating in a machine that, as never

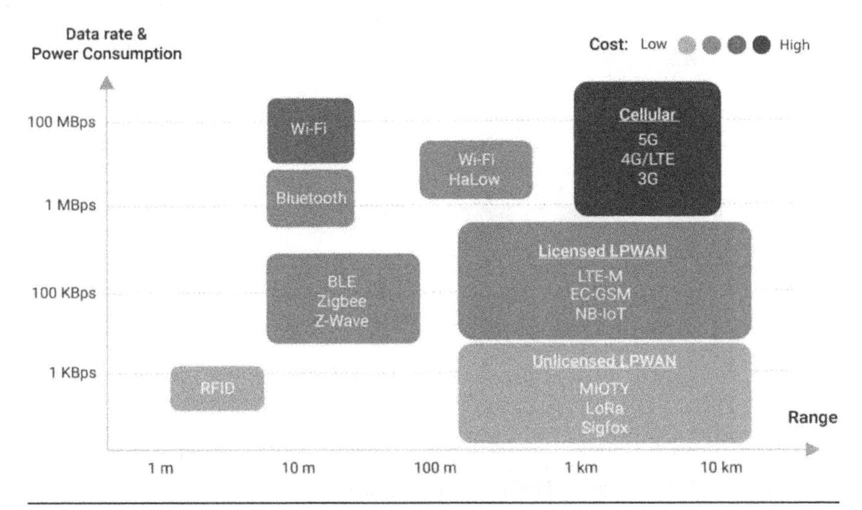

Figure 6.6 Statistics of Power Consumption of IoT devices

before, can track, capture, evaluate, share, and provide new insights. Both these experiences then help guide strategic decisions for businesses that are smarter and quicker.

Technological changes in our daily lives are indispensable in this current era. With the help of emerging new technologies, our lives will become smarter and smarter, alongwith the industrial world. In industrial technology, Industry 4.0 is the latest trend in automation and data sharing. Cyber-physical networks, the IoT and cloud computing, are included (Figure 6.6).

By implementing the IoT in manufacturing industries, a massive change on the quality, quantity, and rate of work per hour will take place. Large numbers of industrial applications are involved in the process of implementing smart manufacturing. Some of the major industrial use cases are discussed below:

- Smart metering
- Assets tracking
- Digital twins
- Smart agriculture
- Smart automobile manufacturing
- Smart farming
- Fleet management
- Livestock management
- Connected vehicles

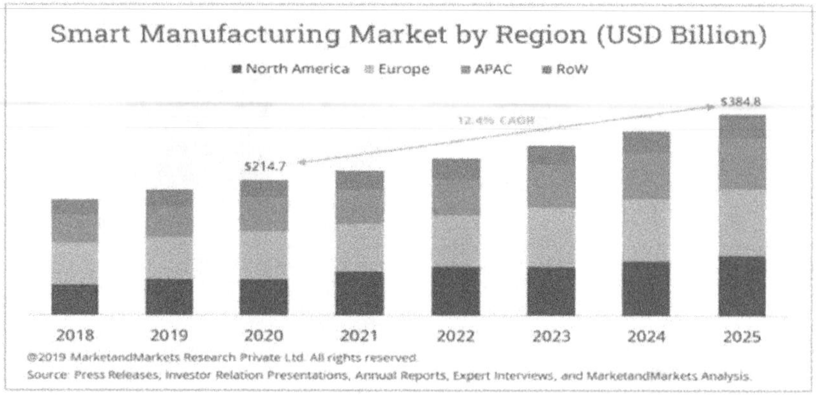

Figure 6.7 Smart Manufacturing Market by Other Regions

What Is Smart Manufacturing?

Generally, we are familiar with the word manufacturing industry, which means an industry or from which a product is produced or made on a large scale. Our traditional manufacturing process involves large number of steps, from choosing the raw materials to packaging the desired product. All this process needs large mechanical machines and man power. Later, computer devices are also involved in manufacturing.

Nowadays, this kind of manufacturing process is getting older and boring. Thus the world is looking forward and is intent toward the new processes of manufacturing, which perfectly suit the hustle and bustle of day-to-day life.

"Smart manufacturing" is the solution for all the problems. By implementing smart manufacturing, most of the problems get solved, and there will be a massive change in the development of the industry in all aspects (Figure 6.7).

In future, smart manufacturing will have a great impact on the manufacturing industries. According to the IDC research, by 2025, the improvement in operational devices by IoT applications could be more than $470 billion per year.

By the following industrial use cases, we can easily implement smart manufacturing in any field and make the industrial world much smarter.

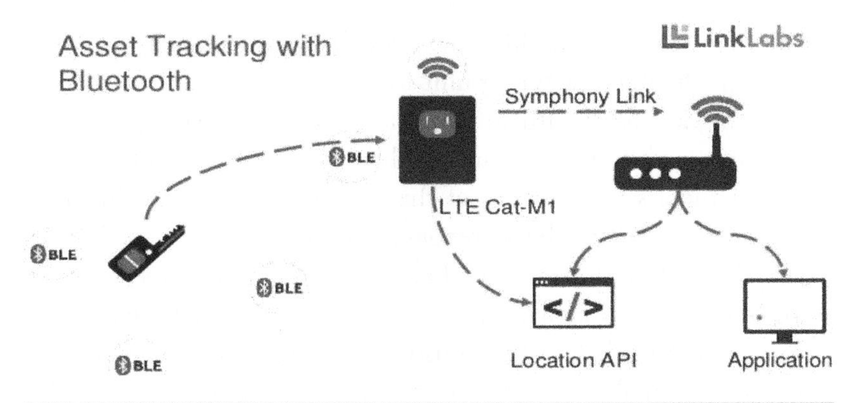

Figure 6.8 Asset Tracking and Locations of IoT

Assets Tracking

Tracking assets refer to the way physical assets are tracked. This can be done either by scanning barcode labels attached to the assets or by using tags that broadcast their location, such as GPS, RFID, or BLE. This technique can also be used to monitor an individual or animal (endangered species).

Assets tracking is a systematic approach that allows the company to track and manage its assets. The best way to improve the management of assets is RFID (radio frequency identification) technology. RFID tags transfer data from an RFID tag to a reader using electromagnetic fields. Listeb below are the systematic approach or steps in an industry to set up asset tracking:

- First, determine a reliable person or team to be responsible for the assets.
- The team would figure out the life cycle of those assets.
- Routine tracking of those assets are mandatory.
- By understanding the functions of the assets in particular operations, we can implement the automated asset tracking (Figure 6.8).

Industrial Example

In a manufacturing industry, large number of assets are involved in the process. So it is essential to track those assets for effective monitoring, for example, lift truck, which is used to carry the products from the manufacturing unit to the warehouse.

If the truck is fixed with an RFID tag, the RFID reader, placed at the exit of the unit, will read and send the data to cloud and software whenever the asset moves out of the manufacturing unit. Similarly, when the asset enters the warehouse, the reader placed at the entrance will read the data and send to the cloud. All the collected data in a cloud system are analyzed, which helps in improving the process.

Assets tracking helps the industry to improve the maintenance, accuracy, accountability, and customer services, and also to lower the administrative expenses. By implementing the assets tracking, we can improve the planning for further growth of the industry.

Given below is the list of top sevenasset management softwares:.

1. ServiceNowITSM
2. Nifty
3. Monday.com
4. Softchoice IT Asset Management
5. NinjaRMM
6. Freshservice
7. Spiceworks IT Asset Management Software

BlackRock, with managed assets amounting to US$6.52 trillion, was the largest asset manager globally in 2019.

Digital Twins

A digital twin may be a replica of a physical entity that is living or not living. It's the Virtual Information Construct group. The potential or physical product of production from the micro atomic level to the macro geometric level is fully described. We can gather any merchandise information from its digital twin as accurately as by physically inspecting the merchandize to obtain the data. It's a virtual, dynamic representation of an object.

Later, the concept was divided into the following types:

□ =>the digital twin prototype (DTP)

□ =>the digital twin instance (DTI)

□ =>the digital twin aggregate (DTA)

In order to understand a physical product, the DTP consists of designs, analyses, and processes. There is a DTP before a physical product exists.

The DTI is the digital twin of each individual instance of the merchandise.

The DTA is an aggregation of DTIs whose data and knowledge will be used for physical product questioning, forecasting, and learning.

The digital twin is supposed to be an up-to-date and accurate copy of the physical object's properties and states, including shape, position, gesture, status, and motion. Simply digital twin means the connectivity between the physical and virtual world. These are connected by the bridge called data. The information would be obtained by using thousands of sensors from a physical device. All these data are then converted to virtual replica. The continuous flow of real-time data from physical entity to its virtual replica merges the truth and virtual. Digital twin is a platform that brings all the experts together providing powerful analysis, insight, and diagnostics. By using digital twins, we can gain more information about the efficiency, safety, performance, and other physical entity at an early stage.

Industrial Example

If we want to know the efficiency of a windmill in power production, we usually make a prototype of that and analyze the things. Then we create a model of the windmill to examine the details of all criteria. These are the methods we have been following for a long time. But nowadays, there is no need for making the prototype and the model. Digital twins make all the things easier. We can extract any necessary information of the windmill by its digital twin. Not only the efficiency, but also the safety risks of an aircraft engine can be known by making a digital twin of it.

Advantages of Digital Twins

Digital twins are boon with the product idea. Their growth will be a continuous process in product development. Thet also predict errors in products, serve as a virtual template during assembly, and remain

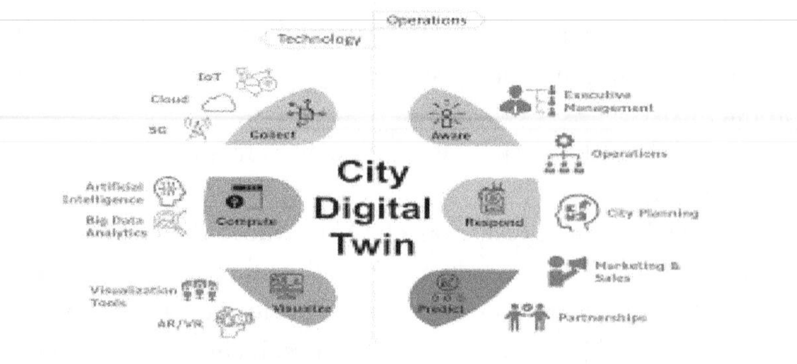

Figure 6.9 Technologies and Operations of Digital Twins

connected for the entire life cycle. They protect reputation and prevent financial cost (Figure 6.9).

Further examples of industry applications are:

- Aircraft engines
- Wind turbines
- HVAC control systems
- Locomotives
- Buildings
- Utilities (electric, gas, water, and wastewater networks)
- Large structures, e.g., offshore platforms, offshore vessels, etc.

Smart Metering

Nowadays, the demand for electricity is increasing sharply due to a number of developmental processes. Maintaining the electricity properly is becoming the greatest challenge. Electricity leakages, tamping, and theft are the major issues in maintaining the bills. We also need manpower to take reading from the electric meters of the users. It is a time-taking process and sometimes the readings taken by the EB person may be wrong or inaccurate, which may lead to extra-charged bills for the users. To eliminate all these circumstances, a new technology has been emerging with the help of IoT called *smart metering* technology. Smart metering is one of the most popular domains of IoT. Smart metering offers advanced metering infrastructure (AMI) via smart meters. It is the smart monitoring and reading of resource usages not only for electricity, but also for water and gas. We can get a reading from a machine, rooms, and indoor or outdoor environments. Once we install the smart meter, the respective meter readings will automatically be sent to the energy supplier, which means that we will receive accurate, not estimated, bills. As all smart meters are certified by the Office for Product Safety & Standards, you can be sure your smart meter is accurate. Smart meters use a secure national communication network (called the DCC) to automatically and wirelessly send your actual energy usage to your supplier. Smart meters send only the reading of the meter to your supplier; your personal details, such as name, contact, and bank details, are neither stored nor transmitted to the internet by smart meters. Smart meters don't need the internet to communicate. Instead, they work by using two wireless networks:

1. HAN (home area network)
2. WAN (wide area network)

These networks are very secure and similar to the mobile communication used to send and receive data. It nearly takes 3 hours for the installation of smart meters for both gas and electricity. If we wish to set installation for any one, it will take nearly 60 to 90 minutes and the gas and power must be switched off for a certain while after the installation is complete. Smart meters are not only used for providing accurate bills of resources, but also provide daily usage details of all the resources with estimated cost; thus, we can reduce and be aware of our daily usage with the help of in-home display. In-home display or IHD is a simple electric device which has a touch screen display. It can be paired with our smart meter and it showcases the needed real-time usage data of our resources in the display.

Advantages of Smart Metering System

- Eliminates the estimated manual monthly meter readings
- Monitors the real-time electric usage data
- Enables the efficient use of power resources
- Provides responsive data for balancing electric loads while reducing blackouts
- Enables dynamic and accurate pricing

- Avoids the capital expense of building new power plants
- Helps to optimize and know the profit with existing resources.

Apart from the above-mentioned advantages, smart metering does many more great things in a smarter way.

Disadvantages of Smart Metering System

- Most smart meters turn dumb whenever we change the supplier.
- Poor network signal prevents the smart meter from working.
- Smart meter stops sending readings due to poor signal, which leads to delayed bills.
- The terms in new smart monitor are hard to understand by elders and vulnerable people.
- Although smart meters do not send our personal details, it poses a risk to security.
- Existing meters are hard to access.

All these mentioned disadvantages stand like a barrier to implement smart metering systems all over the world.

Smart Farming

As the world population increases drastically, the needs of the people also increase simultaneously. Industrial revolutions had played a major role in the transformation of traditional methods to modern solutions. From the past three industrial revolutions, we have made the industrial fields to work in a faster and smarter way by implementing the modern technologies. But in agriculture, we are not able to achieve the major changes like those in large industrial factories. Through the IoT, we can achieve the greatest changes in the agricultural fields by smart farming (Figure 6.10).

Using various types of sensors and IoT devices for monitoring and maintaining the data about the crops and environment is known as smart farming. By this smart farming methods, farmers

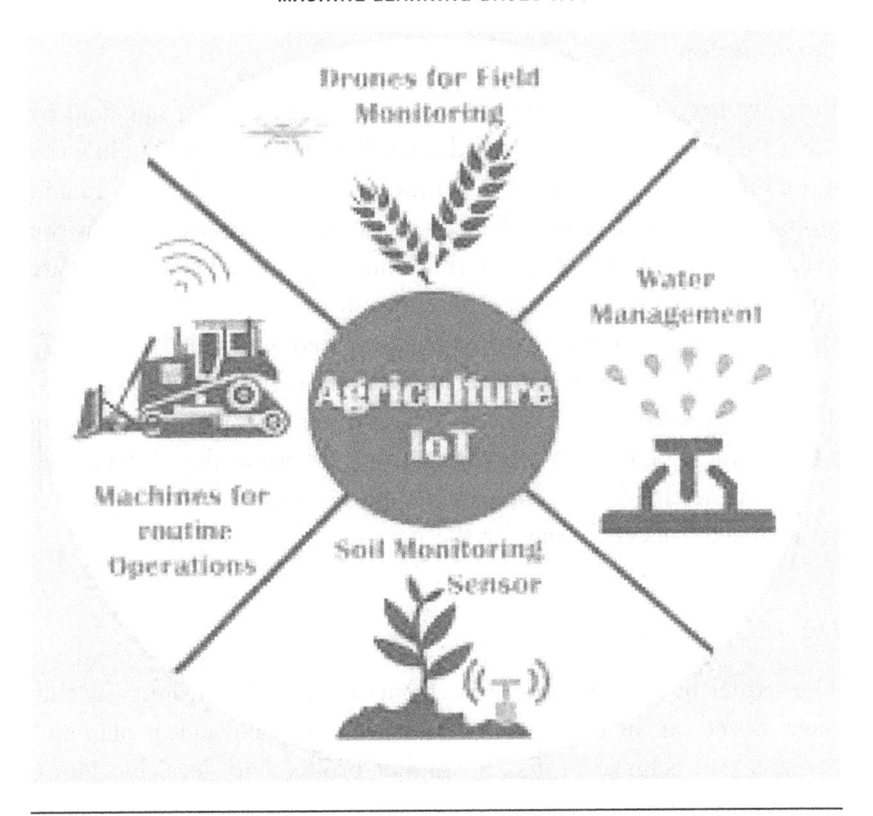

Figure 6.10 Agriculture IoT

will come to know the exact details of their crop at anytime and any where. They will be well-informed about the progressed data of their crops. The farmers can get the details like intensity of sunlight, weather conditions, humidity, temperature, and moisture of the soil, thereby automating the irrigating system in their fields.

How Smart Farming Can Be Done?

There are three major steps involved in the smart farming method:

1. Data collection
2. Data analytics
3. Informed action and planning

Data Collection

First, we need to collect the necessary real-time data of our field by using various types of sensors and other IoT devices. It will help us be more informed about the field. So the sensors must be placed in and around the field at some selected locations, thereby we can get the overall mixed average data of the land. All the collected data are uploaded into a cloud storage for future use.

All the collected data have to be compared and run through different prediction models to get a clear strategic plan for farmers. These analyzed data would be very helpful for the farmers to achieve an optimal growth and yield of the crops. These analyzed data, also stored in the cloud for future use, would be compared with the next set of newly collected data for the next season.

Informed Action and Planning

The farmer have received all the essential actionable insights by this stage. Based on the mentioned factors, he can make a clear plan and decision and achieve profit and higher productivity level by taking appropriate actions.

By following the above-mentioned steps with the help of sensors and other IoT devices, massive revolutionary change can be achieved in agriculture.

Things to Be Considered before Implementing Smart Farming

1. **Hardware**
   We would like to configure the sensors for your device to create an IoT solution for agriculture. Our choice will depend on the data categories we want to collect and thus the purpose of our solution. The standard of selected sensors is, in any case, crucial to the success of the product. The accuracy of the data collected and its reliability will depend on it.
2. **The Brain**
   The core of each smart agriculture solution should be data analytics. If you can't add it up,the collected data itself will be of

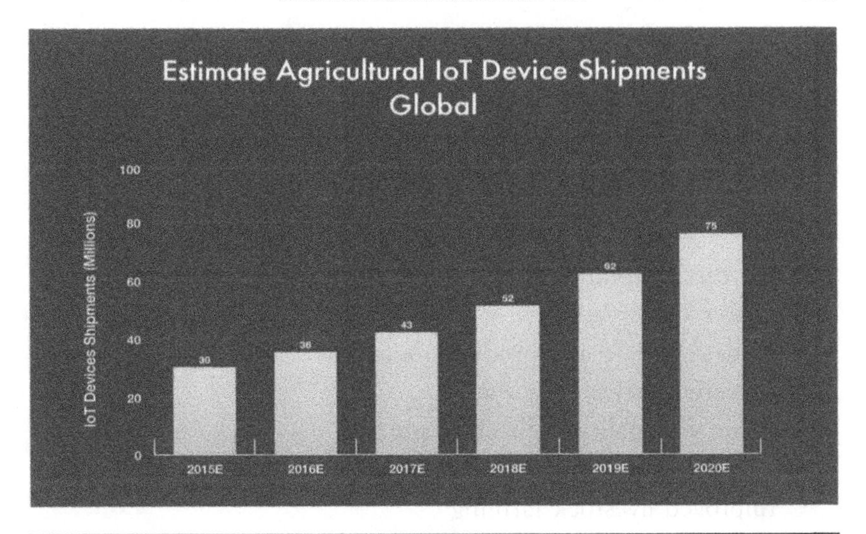

Figure 6.11 Estimation of IoT Device Shipments

little help. Thus, you want to have powerful data analytics capabilities and apply predictive algorithms and ML to support the collected data in order to get actionable insights.

3. **Maintenance**
   Maintenance of your hardware could be a primary challenge for IoT products in agriculture, as the sensors we used in your farm for data collection can easily be damaged. Thus, you want to make sure the hardware is durable and easy to keep up with. Otherwise, more often than you'd like, you'll have to replace your sensors.
4. **Mobility**
   To be used within the field, smart farming applications should be tailored. A business owner or farm manager should be able to access the data via a sensitive phone or PC on site or remotely. In addition, each connected device should be independent and have adequate wireless range to communicate with the opposite devices and transfer data to the central server.
5. **Infrastructure**
   To check that your application for smart farming is doing well. We'll be sure it can handle the load of data. A solid internal infrastructure is what you would need. In addition, the internal networks must be stable. If you have not protected your device,

then stealing your records, or even taking charge of your autonomous tractors, would be very easy for others (Figure 6.11).

Advantages of Smart Farming:

- Increased production
- Water conservation
- Real-time data and production insight
- Lowered operation costs
- Increased quality of production
- Accurate farm and field evaluation
- Improved livestock farming
- Reduced environmental footprint

Livestock Management Using IoT

In crop production, the use of livestock and its sub-product manure is significant. Livestock is an energy source that provides animal power for draught, while manure improves soil structure and fertility, as well as water retention. Enhancing energy and nutrient cycling is environmentally friendly for both uses. Farmers can monitor their livestock very closely by using IoT in the field of livestock management.

Wearables play the vital role in tracking each and every movement of animal. Wearables with connected sensor can track every real-time data and send it to the cloud platform. By using this technology, farmers can improve the standard of livestock breeding and also improve their financial economy in a good manner.

Monitoring the Health of Cattle

Farmers in India suffer a loss of nearly ₹20,000 crore per year due to the adverse effect of animal illness that leads to death. IoT gives farmers a clear solution to those crises and improves better health care in livestock management. IoT offers the farmers a low-cost and low-bandwidth sensor-connected wearables to track the health of cattle. The sensors in the wearables can monitor the blood pressure, heart rate, respiratory rate, digestion, temperature, and other vitals that alert a farmer at the first sign of illness (Figure 6.12).

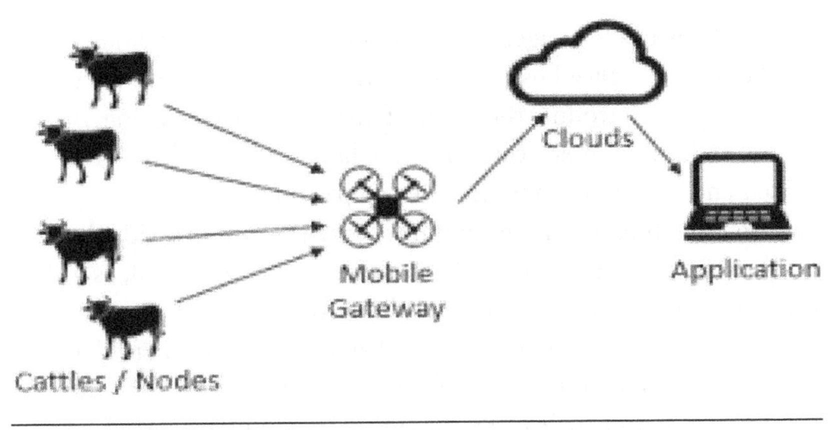

Figure 6.12 IoT in Samrt Farming

Generally, cattle may easily affect disease when they are in a herd. It may lead to enormous loss by spreading to every individual. IoT can easily detect those diseases in a herd and allow the farmer to find out the cattle with chance of disease and isolate it to avoid spreading. IoT also detects the illness of the individual cattle due to feed issue or other parameters and alerts farmers before it becomes critical. For example, a cow's temperature could rise if it is affected by any disease; a small temperature rise rather than normal is enough for IoT to alert the farmer before there is a change in the behavior of the cattle.

Monitoring Reproductive Cycle and Calving

IoT can also be very useful in the reproductive cycle. Before calving, there must be slight rise in the temperature of cow before 8 hours. IoT can alert the farmers about the calving very before, thus increasing the efficiency and productivity of cattle farming. IoT can also inform the farmer when the cow needs to breed and when it is ready for artificial insemination etc. By having these data, a farmer can easily avoid various losses in a herd. If the cow is in a farm during the time of calving, IoT informs the farmer whether she started calving and sends the location of the cow, thus protecting the new born calf from other problems.

Location tracking of cattle in a herd gives enormous benefits to the farmer. Even it is an important feature in health and reproductive cycle of livestock management. The location of an individual cattle and herd can be traced using connected sensors fixed in the wearables which is placed around the neck of the cattle. By tracking the location of herd farmers can come to know, where their cattle are grazed and they also know the area of pastureland.

Farmers can establish and optimize grazing patterns with the data provided from the tracking of the movement of each animal, as well as the migration of the herd. It also allows the farmer to find the sick animal and allow it to be more handled and to isolate the herd that goes into heat from the human form. Overall, it makes it easy for farmers to handle and control their cattle that graze on multiple acres of land.

Fine-Tuning Feeding

Fine-tune feeding is the process of giving a perfect diet to the cattle; it contains all the essential nutrients necessary for the proper growth and lactation of the cattle. By tracking the location of the grazing area, we can easily calculate how frequently they feed. We also come to know about the grazing pattern of the herd, thereby allowing the farmer to feed his herd in an appropriate proportion. By monitoring the specific behavior of the herd such as grazing, socializing, lying down, and chewing the cud, farmer can avoid wasting food or overfeeding the cattle, knowing the frequency of those activities.

Maximizing Milking

IoT also gives dairy farmers a fantastic operation. Robots will literally increase the workload of farmers by reminding farmers when the cow needs to be milked by installing the IoT on a farm, which will lead to increased milking sessions. It can also control any person on a farm's milking speed, quality and quantity of milk produced, quantity of feed taken, and number of steps taken every day. From that data, by deciding the required complement for each person and paying more attention to the cow that produces high quality and quantity of milk, farmers can improve milking. As a result, lactation frequently raises the farm's financial profits at the same time.

Ultimately, farmers are the one to take care of their herds, but IoT can optimise and simplify many major processes in herd farming. IoT easily monitor the location, fertility, health condition, temperature, digestion, movements, grazing patterns, lactation and calving cycle of each and every individual in the farm. By knowing all the data farmers can ultimately increase the productivity, efficiency and revenue of their farm in a progressive manner.

Undoubtedly, apart from agriculture herd, farming has also contributed a huge role in satisfying the needs of people. Implementing the IoT in livestock management will be very effective for the improvement in the life cycle of cattle, as well as human being.

Connected Vehicles

Connected vehicles are one of the most promising products of IoT. Connected vehicles are simply a revolutionary thing at present. It is an upgrade version of ordinary traditional type of vehicle. Connected vehicles can communicate bidirectionally with other systems, both outside and inside the vehicle. Connected vehicles can share the internet access and data with other vehicles for the efficient driving and safety purpose. By IoT vehicle, we can communicate with each other and understand each and every movement of other vehicles (Figure 6.13).

Connected vehicles can share the data all around the vehicle (360°). They collect data from nearby or passing vehicle and respond with required action automatically. Prospectively connected vehicles are able to connect using "dedicated short-range communication

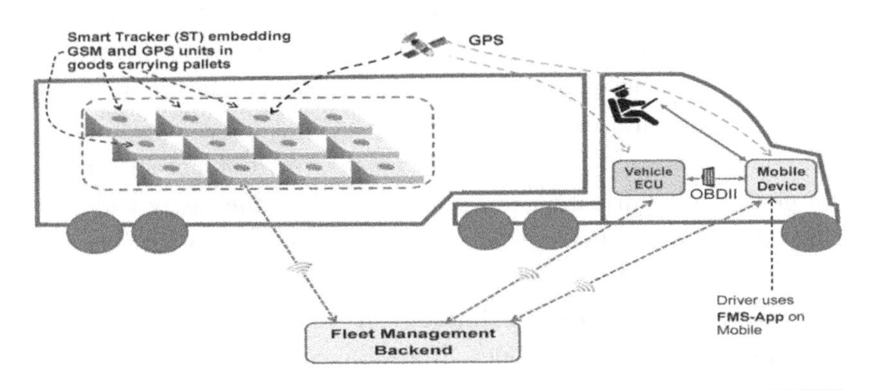

Figure 6.13 Smart Tracking

(DSRC)" radios. They operate in FCC-granted 5.9-GHz band. They have very low latency. Connected vehicles were introduced in early 1996 for the safety purposes, but now they are used for various purposes. Variety of connectivity and usage are enhanced in every upgrade new versions.

Following are the five different types of connectivity available for different purposes:

- **V2V – vehicle to vehicle**
- **V2I – vehicle to infrastructure**
- **V2C – vehicle to cloud system**
- **V2P – vehicle to pedestrian**
- **V2X – vehicle to everything**

Vehicle to vehicle technology offers exchanging the information (data) between the vehicles around it. It shares the information about the current real-time position and speed of the vehicle with accuracy. It is used for the purpose of safety and it avoids unnecessary traffic convention.

Vehicle to infrastructure technology offers the communication of vehicle with the infrastructure of the vehicle. It shares the information generated by the vehicle about the infrastructure to the driver. This V2I technology is used to communicate the information about safety, mobility, and environment information, such as weather conditions, wind speed, and temperature to the driver.

Vehicle to cloud technology uses to share the information about the vehicle to the cloud system. This allows the vehicle to use information from other cloud system of other industries, smart home, etc.

Vehicle to pedestrian technology provides a connection between vehicle to the infrastructure, environment, and other vehicle. This technology generates information about the environment and shares the generated information to other vehicles for the safety and mobility. Thus this technology provides a communication between vehicle and pedestrians on the road.

Vehicle to everything technology is an ultimate goal of connectivity of connected vehicles. It easily enhances communication with infrastructure, environment, pedestrian, and other vehicles. This technology also offers the connection between cars, rails, ships, and airplane. Thus, it does everything in a well precise manner.

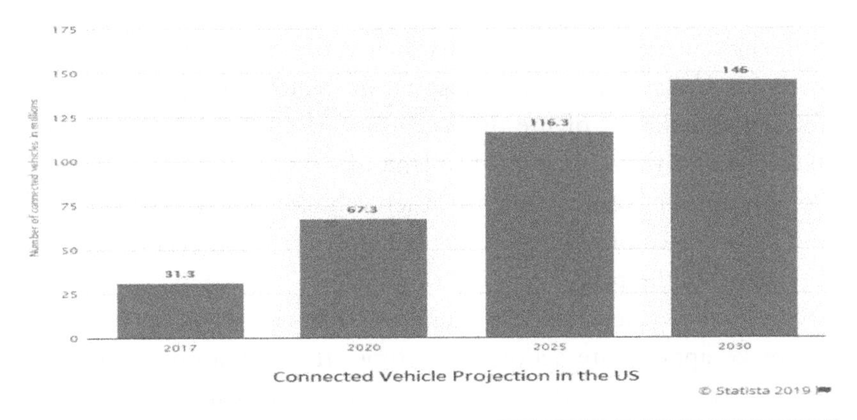

Figure 6.14 Connected Vehicle Prediction in US

Connected vehicles primarily use 3G/4G data connection for sharing the information and communications between vehicle and mobile devices, smart assistants, and websites all over the world. It is estimated that over 28.5 million units of connected cars were sold worldwide in 2019 (Figure 6.14).

- Improves safety on road
- Avoids traffic congestion
- Is Eco friendly
- Provides Urban and suburban mobility
- Avoids accidents
- Provides Infotainment
- Enables enhanced communication
- Emergency calling

Out of these above-mentioned points, connected vehicles have many advantages on real-time applications.

Connected vehicles are going to make a great change on normal roads and traffic. Our new generation is going to experience new things on roads. They connects everything about the vehicle to our hand. We can get all the real-time data about the vehicle, such as speed, quantity of the fuel in the fuel tank, air pressure in the wheel, environmental changes, etc.

If the vehicle is supposed to be met with an accident, the cloud system automatically sents a emergency call to a common center with the current location. This helps the driver to get quick emergency help from nearby hospital or health center, thereby increasing the chance of life of road users. If any accident or repair takes place to the vehicle on the lane of driving, the vehicle sends the appropriate information to other vehicles riding on the same lane before it comes closer. Thus it helps other vehicles to change other lane of the road.

Connected vehicles also communicate with signals on the junction and make appropriate actions with time. If a physically challenged person wants to cross the road, the information will be sent to the vehicle crossing the road so that it can be stopped and the person can cross the crossing. The same information is also transferred to all vehicles moving behind the first one thus avoiding the unwanted accident and traffic. Connected vehicles will improve the standard of driving and the safety factor from all sides. They contribute more to the goodness of the society.

Conclusion

The Internet of things (IoT) is awaiting to give a new revolutionary way of lifestyle to us. Many of us get panic or wondered, when we hear the words such as Internet of Things, artificial intelligence, machine learning, data analysis, block chain, cloud computing etc. We think that these technologies are very complicated, but in reality, they are the simplest things that are going to make ultimate changes in our day-to-day life, if only we are ready to shake our hands with them. It is expected that in the year of 2020, nearly over 50 million devices are going to be connected with IoT in worldwide. Industry 4.0 is one of the ultimate goals of this decade and IoT playing a vital role in the implementation of industry 4.0. IoT has many industrial applications, as we discussed above. They are the key way to the industrial revolution. All the routine ways of traditional manufacturing are digitalized by smart manufacturing. The standard of all manufacturing industries will reach the peak. Thus smart metering, smart agriculture, fleet management, logistics maintenance, digital twins, cloud computing, assets tracking, connected vehicles, and other number of new things through IoT pave a new version of lifestyle for human being and machines. IoT has some drawbacks and disadvantages, but when we compare with benefits, it looks like a piece of dust in the sky. IoT and IIoT are never going to end simply; they are emerging as new and upgrading every day by new technologies with the help of various types of sensors, tracking devices, RFID tags, computing software, etc.

So make a bond with IoT and make everything you want within finger tips. IoT is never be a period, it will always remain as a comma.

References

[1] Ismail, Yasser, "Introductory Chapter: Internet of Things (IoT) Importance and Its Applications," in Yasser Ismail (ed.), *Internet of Things (IoT) for Automated and Smart Applications*. IntechOpen.com, November 27, 2019, doi: 10.5772/intechopen.90022.

[2] Tracy, Phillip, "The Top 5 Industrial IoT Use Cases," 5 minute read, IBM.com, April 19, 2017, https://www.ibm.com/blogs/internet-of-things/top-5-industrial-iot-use-cases

[3] DdGubbia, J., Buyyab, R., Marusic, S., and Palaniswami, M., "Internet of Things (IoT): A vision, architectural elements, and future directions," *Future Generation Computer Systems*, vol. 29, September 2013.

[4] EeFlexeye, "Lord of the Things: Why Identity, Visibility and Intelligence are the Key to Unlocking the Value of IoT," 2014, https://coe.flexeyetech.com/.

[5] Xu, L., He, W., and Li, S., "Internet of Things in Industries: A Survey," *IEEE Transactions on Industrial Informatics*, vol. 10, no. 4, November 2014, pp. 2233–2243.

[6] *IEEE Internet of Things Journal*, http://standards.ieee.org/innovate/iot/

[7] Gilchrist, Alasdair, *Professional and Applied Computing*, eBook Packages, Berkeley, CA: Apress, 2016, https://doi.org/10.1007/978-1-4842-2047-4.

[8] Elangovan, Uthayan, *Smart Automation to Smart Manufacturing: Industrial Internet of Things*, Manufacturing and Processes Collection, New York: Momentum Press, 2019, ISBN:1949449262,9781949449266.

[illegible] company with detectors, or looks like a piece of dust in the sky. In IIoT, any uncertainties and anomaly, they are emerging as new and [illegible] [illegible] of [illegible] [illegible]

References

[1] [illegible]
[2] [illegible]
[3] [illegible]
[4] [illegible]
[5] [illegible]
[6] [illegible]
[7] [illegible]
[8] [illegible]

7

Employee Turnover Prediction Using Single Voting Model

R. VALARMATHI[1], M. UMADEVI[2], AND T. SHEELA[3]

[1]*Department of Computer Science and Engineering, Sri Sairam Engineering College, Chennai*
[2]*Department of Computer Science and Engineering, SRM Institute of Science and Technology, Chennai*
[3]*Department of Information Technology, Sri Sairam Engineering College, Chennai*

Contents

DOI: 10.1201/9781003119838-7

Introduction

Switching of employees working in a company to another company is called attrition. Employees change company due to lack of monetary benefits, job pressure, work–life balance, lack of promotional opportunities, etc. The employee attrition rate differs with age group, gender, marital status, department of work, and job role. When the attrition rate goes high, companies and employers face various problems. Selecting the proper employees with matching educational qualification and work expertise from market is a challenge. Recruiting freshers or replacements of existing employees need training and time to get the expertise in the domain, tools used, documents to be prepared, etc. Client service gets affected during the transition period. Service quality goes down due to the newcomers trying their best in the new work areas. Due to this, some clients offer the work assigned to some other companies with lesser attrition rate.

Role of Supervisors

Role of supervisors in boosting the employee morale plays the main part. Supervisors have to recommend the right candidates for promotion and onsite works and rate the employees correctly in annual appraisals. New employees learn the work ethics and get trainings from the supervisor mainly. Depending on employee preferences and personal commitments, supervisors can recommend or decide the department transfers, work location transfers, and working hour change.

Role of HR Department

The Human Resources (HR) department works as a brain center of any company. HR department has to devise right company policies and take decisions so that the employees get the work–life balance and correct annual remunerations. HR plays the main role in recruiting right candidates, when vacancies arise, so that the supervisors' burden and time spent on training get reduced. Companies' annual leave policies, working hours, and facilities, like transport, health insurance, and provident fund, are decided by the HR department only. Also, recreational programs and employees' engagement facilities are conducted by HR department, which decides the attrition rate in a company.

Related Work

The proposed work chooses the top 10 features using CART analysis. The imbalanced dataset is split into 60% and 40%, thereby reducing the bias and improving the sensitivity and specificity of the ensemble model built [1]. The authors have addressed the employee turnover issue with six different machine learning (ML) algorithms, namely, SVM (RBF Kernel), XGBoost, logistic regression, Naive Bayesian, Random Forest with Depth controlled, and LDA and KNN. As predicted, the algorithms perform with different measures, such as AUC, running time, and memory utilization. The experimental results shows that XGBoost outperforms all other algorithms in terms of accuracy and memory utilization. Logistic regression took a minimum of 52 seconds to run the prediction algorithm [2]. The author demonstrated the prediction of employee turnover by analyzing varying size (small, medium, and large) and complexity of the human resource datasets with 10 different supervised classifiers. The author has provided guidelines for using the statistical methods and recommended gradient boosting for analyzing large datasets, as it takes less time for data preprocessing and ranks the features automatically and reliably [3]. Employee attrition dataset created by IBM Watson predicts the employee turnover [4].

The researchers proposed a weighted quadratic random forest method. The author followed a two-step process. In the first step, the dimensions are reduced and important features are selected. In the second step, *F*-measure is applied on the selected features for each decision tree. Two most important features that influence employee turnover identified by the author are monthly income and overtime [5]. The algorithm lacks in ranking the best features [6–9] in these studies. The researchers studied the impact of pay satisfaction and school achievement on principals and found a negative correlation with principal turnover [10].

The authors investigated the hospital nurse turnover and found job satisfaction, age, and stress to be the major factors that influence the nurse turnover [11]. Rotation forest approach helps in improving individual classifier's accuracy and diversity. Diversity is achieved through features axes and accuracy through principal components [12]. The researchers [13] stated that rotation forest is a better and accurate approach for the real valued attributes compared to SVM, NN, and tree-based ensembles. They evaluated the performance on three different datasets containing about 200 classification problems based on four parameters, such as errors, ROC (receiver operating characteristic), negative likelihood, and balanced error. The Naive Bayes model [14] was employed on HR dataset to predict the attrition. The authors [15] used dynamic partite graph to model the job records.

System Implementation

The system is fragmented into a number of segments for better understanding. The integral segments within the systems are as follows:

- Data preprocessing
- Predicting the employee turnover using logistic regression and Naive Bayes classification using grid search
- Combining the logistic regression and Naive Bayes classification as a single voting model using grid Search to predict the employee turnover by averaging the two models

The workflow is depicted in Figure 7.1.

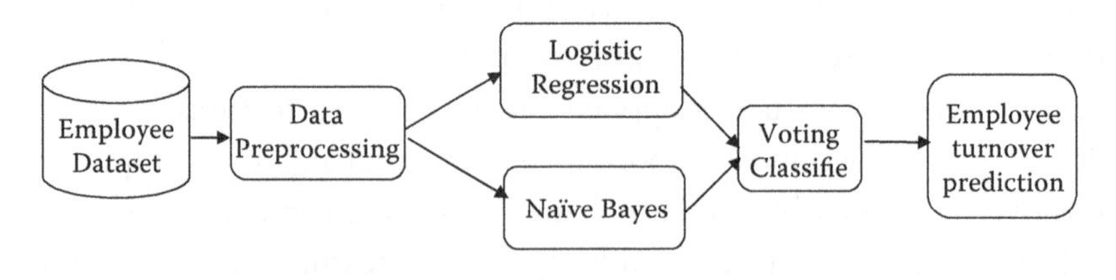

Figure 7.1 System Flow Diagram

Data Preprocessing

The datasets are preprocessed to improve the accuracy of the model. The HR dataset and employee satisfaction level are merged. Label encoding is applied to convert names to numeric.

Materials and Methods Used

Logistic Regression Logistic regression is a binary classification model to find the probability of a certain event existing, such as tail/head, 0/1, or unhealthy/healthy. The binary classification model is a problem with two class values, such as True/False and Buy/Sell. It is a statistical model to predict the categorical dependent variable by finding the value for it using a logit function. In 1972, Nelder and Wedderburn created this method, which is the part of generalized linear model (GLM).

To predict the continuous valued quantities, like price of a house, linear regression model which is a linear function of input values (like the square feet of the house) is used. To predict the discrete valued quantities like predicting the pixel intensities representing a "0" digit or a "1" digit, logistic regression model which is a logistic function of input values is used. So it is a simple classification algorithm for learning to make such decisions. It is one of the supervised ML models. In supervised learning, the response variable (B) is represented in terms of input variable (A) that maps the input to the output.

$$B = \mathrm{f}(A)$$

When the response variable (B) is of a categorical type like "green"/"red," "good"/"bad," then classification model is used, whereas when the response variable (B) is continuously valued like "dollars" or "weight," then linear regression model is used.

This model is used to calculate a value for the response variable with only values like "married"/"unmarried" or "male"/"female" by using the given independent variables. So, it is one type of linear regression model with categorical value for the predictor, which uses log of odds as dependent variable.

Different Categories of Logistic Regression

1. **Binary Logistic Regression**

When the dependent variable (target) has merely two probable outcomes, like "happy" or "sad," then binary logistic regression is used.

2. **Multinomial Logistic Regression**

When the dependent variable (target) has three or more possible responses without ordering like the color that is preferred more, "Red" or "Blue" or "Green," then multinomial logistic regression is used.

3. **Ordinal Logistic Regression**

When the dependent variable (target) has three or more possible responses with ordering, like the grade value of the marks be ("A," "A+," "B," "B+," "F"), then ordinal logistic regression is used.

Deriving Logit Function The response variable Y which is a function of predictor/independent variable X is represented as follows:

$$Y = \alpha + \beta(X) \tag{7.1}$$

where, α is a constant term which will be the probability of the event happening when no other factors are considered. Equation (7.1) is a link function that represents the binary outcome Yes or No. We can create a dummy variable P to indicate if the observation is a Yes or a No (i.e., $P = 1$ / $P = 0$). It is done as follows using the two things:

i. The probability of getting Yes (P)
ii. The probability of getting No ($1 - P$), where $0 \geq P \leq 1$.

The probability value should be positive and to make it positive, we find the exponent value of Equation (7.1) as:

$$P = e^{\alpha+\beta(X)} \tag{7.2}$$

To confine P to $0 \geq P \leq 1$, divide P by the value which is at least one value greater than P:

$$P = \frac{e^{\alpha+\beta(X)}}{1 + e^{\alpha+\beta(X)}} \tag{7.3}$$

By Equation (7.1), $Y = \alpha + \beta\ (X)$ and by combining Equations (7.1), (7.2), and (7.3), we can write P as follows:

$$P = \frac{e^y}{1 + e^y} \tag{7.4}$$

where, P is the Logit function which gives the probability of getting "Yes." Now the probability of getting "No" $1 - P$ is written as follows:

$$1 - P = 1 - \frac{e^y}{1 + e^y} \tag{7.5}$$

By dividing Equations (7.4) and (7.5), we can get

$$\frac{P}{Q} = \frac{P}{1 = P} = \frac{\frac{e^y}{1+e^y}}{1 - \frac{e^y}{1+e^y}} \tag{7.6}$$

By simplifying Equation (7.6), we can obtain

$$\frac{P}{Q} = \frac{P}{1 - P} = e^Y \tag{7.7}$$

Taking logarithmic on Equation (7.7), we can obtain

$$\begin{aligned} \text{Log}(e^Y) &= Log\left(\frac{P}{1-P}\right) \\ \text{Y} &= Log\left(\frac{P}{1-P}\right) \end{aligned} \tag{7.8}$$

By substituting (7.1) in (7.8), we will get

$$Log\left(\frac{P}{1 - P}\right) = \alpha + \beta(\text{X}). \tag{7.9}$$

where $\frac{P}{1-P}$ is the odd ratio.

It is a link function used in logistic regression. The logarithmic value on the dependent variable is used to mode a non-linear association. Whenever the logarithmic value of the odd ratio is positive, the probability of success is always more than 50%.

Performance of Logistic Regression Model The various performance validation parameters of logistic regression model are null deviance and residual deviance, confusion matrix, and AUC-ROC curve.

1. **Null Deviance and Residual Deviance**

Null deviance is the measure of the outcome predicted by the model without considering any input, whereas residual deviance is the measure of the outcome predicted by the model in consideration with the independent variable. Lower the value, better the performance.

2. **Confusion Matrix**

It is the representation of actual class outcome versus predicted class outcome in table form, shown below, which is used to measure other metrics, like accuracy, precision, recall, and F-Measure.

$$\text{Specificity/TN Rate} = \frac{A}{A + B}$$

$$\text{FP Rate} = \frac{B}{A + B}$$

$$\text{Sensitivity/TP Rate} = \frac{D}{C + D}$$

$$\text{FN Rate} = \frac{c}{c + d}$$

Actual class outcome	Predicted class outcome	
	Yes	No
Yes	True Positive – TP (D)	False Negative – FN (C)
No	False Positive – FP (B)	True Negative – TN (A)

TP and TN represent the data items that are correctly predicted, whereas FP and FN represent the data items that are not correctly predicted, i.e., the actual class contradicts with the predicted class.

True Positives (TP) – It shows the acceptable predicted values, i.e., actual class outcome and predicted class outcome have the same value.

True Negatives (TN) – It shows that predicted values are invalid, i.e., actual class outcome and predicted class outcome have different values.

False Positives (FP) and False Negatives (FN) – It shows that predicted values are invalid, i.e., actual class outcome and predicted class outcome have contradicted values.

Accuracy – It is a percentage over the total number of observations that are predicted correctly to the total number of observations. Higher the accuracy, better the model. If the model has symmetric datasets consisting of equal number of false positive values and false negative values, then accuracy is considered as a great measure of performance. It is written in Equation (7.10) as:

$$\text{Accuracy} = \frac{A + D}{A + B + C + D} \tag{7.10}$$

Precision – It is the percentage of the total number of correctly predicted observations to the total number of positively predicted observations. Whenever the precision is high, the false positive rate is low.

$$\text{Precision} = \text{D/B} + \text{D}$$

Recall (Sensitivity) – It is the percentage of the total number of positively predicted observations to all the observations in actual class – yes.

$$\text{Recall} = \text{D/C} + \text{D}$$

F1 Score – The trade-off between precision and recall is represented as F-Score in Equation (7.11). Whenever there is an uneven class distribution, this score is usually more practical than accuracy.

$$\text{F1} - \text{Score} = 2 * (\text{Recall} * \text{Precision})/(\text{Recall} + \text{Precision}) \tag{7.11}$$

3. **ROC Curve (Receiver Operating Characteristic Curve)**

It gives the tradeoff between the true positive and false positive. As the model is concerned about the success rate, assume that the P value is greater than 0.5. ROC curve represents all possible values of P which is higher than 0.5.

Naive Bayes Classification Naive Bayes classification is a an ML algorithm based on Bayes Theorem for predictive analysis. It is a classification model based on conditional probability. The conditional probability is a measure of the probability of an occurrence of an event A after an occurrence of another event B. It calculates the probabilities of discrete words like spam or not, married or unmarried, and so on. These Probabilities are also called likelihoods. It consists of the following two parts:

1. Naive
2. Bayes

Bayes Theorem

It is a theorem for calculating conditional probability of an event. Consider events A and B. Probability of an event A to occur is P(A) and probability of an event B to occur is P(B). Bayes theorem shows the relationship between the probability of an event A before occurring in the event B. Probability of an event A after getting the event B, P(A/B), is as follows:

$$\mathrm{P}\left(\frac{A}{B}\right) = \frac{P(\frac{B}{A}).\ P(A)}{P(B)}, \tag{7.12}$$

where A is hypothesis, B is evidence, P(A/B) is conditional probability of A given B which tells that how often A occurs given that B occurs. It is posterior probability, i.e., probability of the event after the evidence has occurred. P(B/A) is conditional probability of B given A which tells that how often B occurs given that A occurs. P (A) is prior probability, i.e., probability of the event before evidence has occurred.

Dataset

The dataset used for the classification consists of:

- Set of independent variables called input features
- Dependent/output variable

Each attributes in the input features are mutually independent. There is always a strong independence among the features in the feature matrix. Hence, it is called Naive. Also each attributes in the input features have an equal contribution in predicting the output class.

Categories of Naive Bayes Classification

Multinomial Naive Bayes This is the multivariate event model to categorize a text/article into games, news, technical, cine field, etc.

Bernoulli Naive Bayes When the dependent variable (target) has merely two probable outcomes like "Happy" or "sad," the classification model used is Bernoulli Naive Bayes. It is the binomial model as the input independent variables take only binary values "Yes" or "No."

Gaussian Naive Bayes When the independent variables have continuous values rather than discrete values, Gaussian Naive Bayes classification can be used by sampling them using Gaussian distribution or normal distribution.

Benefits of Using Naive Bayes

- Straightforward method and high-speed classification model
- Suitable for multi-class classification
- Effectively works with outsized datasets
- Performs better than the other models, like logistic regression
- A smaller amount of training data is required
- Performance is better for categorical data than continuous data
- Low computation cost
- Most suitable for text classification problems

Areas of Applications

- Superior than other algorithms and so it is used in all kinds of binary classifications
- Naive Bayes along with collaborative filtering are used in recommended systems
- It is also used in disease prediction based on health parameters
- Found its application in face recognition
- Naive Bayes is used in prediction of weather reports based on atmospheric conditions (temp, wind, clouds, humidity, etc.)
- Can be used in real-time prediction
- Can be used in multiclass prediction

Voting Classifier The predictions of multiple classifiers are combined. Multiple classifiers are used to train and validate the dataset. Finally, the predictions are taken from the majority voting. Hard voting and soft voting are the different categories of majority voting.

Hard Voting In hard voting, the final predicted class label will be the label predicted by most of the classifiers. For example, assume there are three different classifiers and two different classes, Dog and Cat. If Dog is predicted by two classifiers and Cat is predicted by one classifier, final prediction Dog will be chosen.

Soft Voting In soft voting, the final predicted class label will be the probabilities of all the classifiers chosen for the problem. For example, take three classifiers M1, M2 and M3 and two classes Dog and Cat. Assume

M1 – >{0.4, 0.7}
M2 – >{0.2, 0.8}
M3 – >{0.7, 0.3}

$$\text{Probability of Class Dog} = 0.33 * 0.4 + 0.33 * 0.2 + 0.33 * 0.7 = 0.429$$

$$\text{Probability of Class Cat} = 0.33 * 0.7 + 0.33 * 0.8 + 0.33 * 0.3 = 0.594$$

The probabilities predicted by the combination of classifier are [42.9%, 59.4%]. The class Cat which is having prediction probability of 59.4% will be chosen. The above is an example of soft voting classifier with equal weights.

Results and Discussions

All the three models are executed in Python 3.0 and scikit-learn Libraries.

In [1]:

```
import numpy as np
import pandas as pd
emp_data=pd.read_csv(r'D:\Research\Employee attrition\hr_data.csv')
```

In [2]:

```
emp_data.head()
```

Out [2]:

	EMPLOYEE_ ID	NUMBER_ PROJECT	AVERAGE_ MONTLY_ HOURS	TIME_SPEND_ COMPANY	WORK_ ACCIDENT	LEFT	PROMOTION_ LAST_ 5YEARS	DEPARTMENT	SALARY
0	1003	2	157	3	0	1	0	sales	low
1	1005	5	262	6	0	1	0	sales	medium

(*Continued*)

	EMPLOYEE_ ID	NUMBER_ PROJECT	AVERAGE_ MONTLY_ HOURS	TIME_SPEND_ COMPANY	WORK_ ACCIDENT	LEFT	PROMOTION_ LAST_ 5YEARS	DEPARTMENT	SALARY
2	1486	7	272	4	0	1	0	sales	medium
3	1038	5	223	5	0	1	0	sales	low
4	1057	2	159	3	0	1	0	sales	low

In [3]:

```
emp_data.shape
```

Out [3]:
(14999, 9)

In [4]:

```
emp_data.size
```

Out [4]:
134991

In [5]:

```
sat_data=pd.read_excel(r'D:\Research\Employee attrition\
   employee_satisfaction_evaluation.xlsx')
```

In [6]:

```
sat_data.head()
```

Out [6]:

	EMPLOYEE #	SATISFACTION_LEVEL	LAST_EVALUATION
0	1003	0.38	0.53
1	1005	0.80	0.86
2	1486	0.11	0.88
3	1038	0.72	0.87
4	1057	0.37	0.52

In [7]:

```
empatt_data= emp_data.set_index('employee_id').join(sat_data.set_index('EMPLOYEE #'))
```

In [8]:

```
empatt_data.describe()
```

Out [8]:

	EMPLOYEE_ID	NUMBER_ PROJECT	AVERAGE_ MONTLY_ HOURS	TIME_SPEND_ COMPANY	WORK_ ACCIDENT	LEFT	PROMOTION_ LAST_5YEARS	SATISFACTION_ LEVEL	LAST_ EVALUATION
count	14999.0000	14,999.000	14,999.0000	14999.0000	14999.0000	14999.0000	14999.0000	14972.0000	14972.0000
mean	45424.6275	3.803054	201.050337	3.498233	0.144610	0.238083	0.021268	0.612830	0.716125
std	25915.9001	1.232592	49.943099	1.460136	0.351719	0.425924	0.144281	0.248714	0.171138
min	1003.00000	2.000000	96.000000	2.000000	0.000000	0.000000	0.000000	0.090000	0.360000
25%	22872.5000	3.000000	156.000000	3.000000	0.000000	0.000000	0.000000	0.440000	0.560000
50%	45448.00	4.000000	200.000000	3.000000	0.000000	0.000000	0.000000	0.640000	0.720000
75%	67480.500000	5.000000	245.000000	4.000000	0.000000	0.000000	0.000000	0.820000	0.870000
max	99815.000000	7.000000	310.000000	10.000000	1.000000	1.000000	1.000000	1.000000	1.000000

In [9]:

```
from sklearn.preprocessing import LabelEncoder
lab=LabelEncoder()
n=lab.fit_transform(empatt_data['salary'])
```

In [10]:

```
empatt_data['salary_num']=n
empatt_data.drop(['salary'],axis=1,inplace=True)
z=lab.fit_transform(empatt_data['department'])
empatt_data.drop(['department'],axis=1,inplace=True)
X=empatt_data.drop(['left'],axis=1)
```

In [11]:

```
X.head()
```

Out [11]:

	NUMBER_ PROJECT	AVERAGE_ MONTLY_ HOURS	TIME_ SPEND_ COMPANY	WORK_ ACCIDENT	PROMOTION_ LAST_ 5YEARS	SATISFACTION_ LEVEL	LAST_ EVALUATION	SALARY_NUM
0	2	157	3	0	0	0.38	0.53	1
1	5	262	6	0	0	0.80	0.86	2
2	7	272	4	0	0	0.11	0.88	2
3	5	223	5	0	0	0.72	0.87	1
4	2	159	3	0	0	0.37	0.52	1

In [12]:

```
y=empatt_data['left']
```

In [13]:

```
y.head()
```

Out [13]:

```
01
11
21
31
41
```

Name: left, dtype: int64

In [14]:

```
#import the libraries
import matplotlib.pyplot as plt
%matplotlib inline
import seaborn as sns
from sklearn.preprocessing import StandardScaler
# Split data into Training and Testing Dataset
from sklearn.model_selection import train_test_split
# Data modeling
from sklearn.metrics import confusion_matrix,accuracy_score,roc_curve,classification_report
from sklearn.metrics import roc_auc_score
from sklearn.linear_model import LogisticRegression
from sklearn.naive_bayes import GaussianNB
```

In [15]:

```
from sklearn.model_selection import train_test_split
X_train_data, X_test_data ,y_train_data ,y_test_data =train_test_split (X, y, test_size=0.10,
   random_state=0)
```

Logistic regression has different hyper parameters solver, penalty and C. Different solvers result in different types of performance: penalty helps in regularization and C helps in controling the penalty strength. These parameters help to optimize logistic regression. Logistic regression can be evaluated with different combination of parameters and the best hyper parameters can be found. Grid search is an exhaustive search strategy that considers all parameter combinations while sampling in the search space.

In [16]:

```
#Model 1-Logistic Regression
from sklearn.model_selection import GridSearchCV
from sklearn.linear_model import LogisticRegression
log_reg = LogisticRegression()
params = {'C': [1e-4, 1e-3, 1e-2, 1e-1, 0.5, 1., 5., 10., 15., 20., 25.]}
#Grid Searchwith 10-fold cross validation
log_model1 = GridSearchCV( log_reg , param_grid=params,cv=10, n_jobs=-1)
log_model1.fit(X_test_data, y_test_data)
#The Best Hyper parameters of Logistic Regression
print("Best Hyper Parameters:\n", log_model1.best_params_)
#Prediction
lr_predict= log_model1.predict(X_test_data)
#import metrics
from sklearn import metrics
#Accuracy
lr_acc_score = accuracy_score(y_test_data, lr_predict)
print("Accuracy:",metrics.accuracy_score(lr_predict, y_test_data))
#Confusion Matrix
print("Confusion Matrix:\n",metrics.confusion_matrix (lr_predict, y_test_data))
#Classification Report
print(classification_report (y_test_data,lr_predict))
#ROC Score
print("ROC_AUC_SCORE")
roc_auc_score(y_test,lr_predict*100)
```

Output:

```
Best Hyper Parameters:
{'C': 25.0}
Accuracy: 0.77
Confusion Matrix:
[[1052 256]
[ 89 103]]
              precisionrecallf1-scoresupport
            00.80      0.92 0.86     1141
            10.54      0.29 0.37     359
accuracy                    0.77     1500
macro avg     0.67     0.60 0.62     1500
weighted avg  0.74     0.77 0.74     1500
ROC_AUC_SCORE
Out[16]:
0.6044531625730252
```

The parameter var_smoothing helps to tune GaussianNB

In [17]:

```
#Model 2-Naive Bayes Classifier
from sklearn.naive_bayes import GaussianNB
nb_c = GaussianNB()
nb_params = { 'var_smoothing':np.logspace(0,1,num=100)}
#Grid Search with 10-fold cross validation
nbmodel = GridSearchCV(nb_c, param_grid=nb_params,cv=10, n_jobs=-1)
#Learning
nbmodel.fit(X_test_data ,y_test_data)
#The Best Hyper parameters are
print("Best Hyper Parameters:\n",nbmodel.best_params_)
#Prediction
nbpred=nbmodel.predict(X_test_data)
#Accuracy
nb_acc_score = accuracy_score(y_test_data, nbpred)
print("Accuracy:",metrics.accuracy_score(nbpred, y_test_data))
#Confusion Matrix
print("Confusion Matrix:\n",metrics.confusion_matrix (nbpred, y_test_data))
#Classification Report
print(classification_report(y_test_data, nbpred))
#ROC Score
print("ROC_AUC_SCORE")
roc_auc_score(y_test_data, nbpred*100)
```

Output:

```
Best Hyper Parameters:
{'var_smoothing': 1.0}
Accuracy: 0.7606666666666667
Confusion Matrix:
[[1141 359]
[0 0]]
              precisionrecallf1-scoresupport
0             0.76      1.00 0.86     1141
1             0.00      0.00 0.00     359
accuracy                     0.76     1500
macro avg     0.38      0.50 0.43     1500
weighted avg0.58        0.76 0.66     1500
ROC_AUC_SCORE
Out[17]:
0.5
```

Accuracy of logistic regression (LR) and Naive Bayes (NB) classifier are separately found in the above-mentioned modules. The single voting model that combines the majority voting of LR and NB is shown below:

In [18]:

```
#Model 3-Voting Classifier
from sklearn.ensemble import VotingClassifier
mod1 = LogisticRegression()
mod2 = GaussianNB()
vc = VotingClassifier(estimators=[('lr',mod1),
('nb',mod2)],
voting='soft',n_jobs=-1)
params = {'weights':[[1,2],[1,2]]}
grid_vc = GridSearchCV(param_grid = params,cv=10, estimator=vc)
grid_vc.fit(X_test_data ,y_test_data)
print(grid_vc.best_params_)
grid_predict=grid_vc.predict(X_test_data)
#Accuracy
grid_acc_score = accuracy_score (y_test_data, grid_predict)
print("Accuracy:",metrics.accuracy_score(grid_predict, y_test_data))
#Confusion Matrix
print("Confusion Matrix:\n",metrics.confusion_matrix(grid_predict,y_test_data))
#Classification Report
print(classification_report(y_test_data,grid_predict))
#ROC Score
print("ROC_AUC_SCORE")
roc_auc_score(y_test_data, grid_predict*100)
```

Soft voting is used which produces the output as the average probability of the two models. The combined voting method shows the highest accuracy of 84.93% when compared to LR and NB run separately.

Output:

```
{'weights': [1,2]}
Accuracy: 0.848
Confusion Matrix:
[[1021 108]
[120 251]]
              precisionrecallf1-scoresupport
0             0.90      0.89 0.90    1141
1             0.68      0.70 0.69    359
accuracy                     0.85    1500
macro avg     0.79      0.80 0.79    1500
weighted avg0.85        0.85 0.85    1500
ROC_AUC_SCORE
Out[18]:
0.7969967213434924
```

Visualization of true positive rate and false positive rate:

In [19]:

```
#Reciever Operating Characterstic Curve
log_reg_fpr,log_reg_tpr, log_reg_threshold = roc_curve(y_test_data,lr_predict)
nb_c_fpr, nb_c_tpr, nb_c_threshold = roc_curve(y_test_data, nbpred)
rf_fpr,rf_tpr,rf_threshold = roc_curve(y_test_data, grid_predict)
sns.set_style('whitegrid')
plt.figure(figsize=(8,5))
plt.title('Reciever Operating Characterstic Curve')
plt.plot(log_reg_fpr,log_reg_tpr,label='Logistic Regression')
plt.plot(nb_c_fpr,nb_c_tpr,label='Naive Bayes')
plt.plot(rf_fpr,rf_tpr,label='VC')
plt.plot([0,1],ls='--')
plt.plot([0,0],[1,0],c='.5')
plt.plot([1,1],c='.5')
plt.ylabel('True positive rate')
plt.xlabel('False positive rate')
plt.legend()
plt.show()
```

Figure 7.2 shows the ROC-AUC curve obtained for the three models.

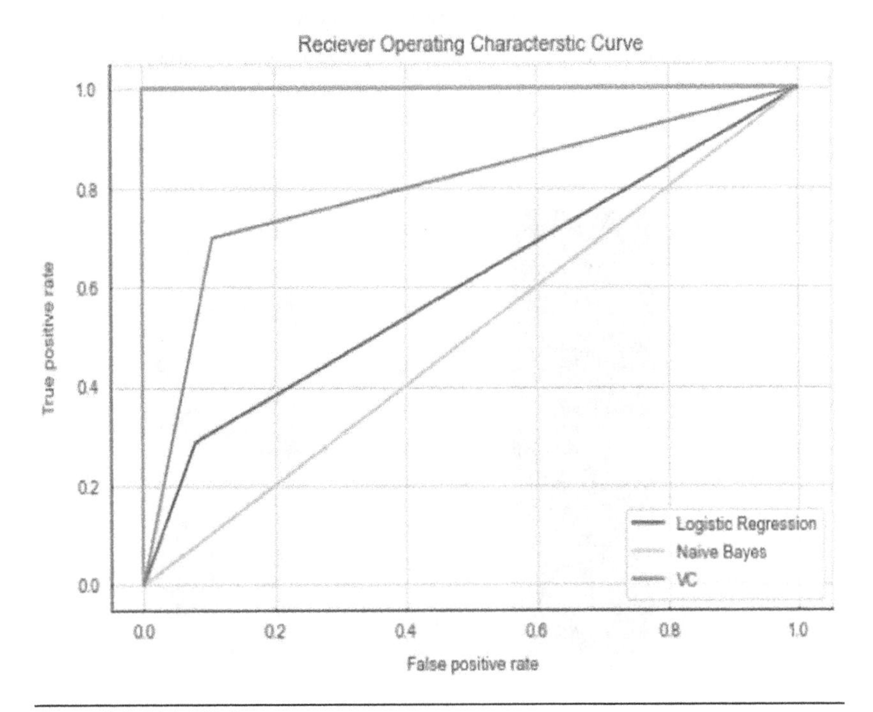

Figure 7.2 ROC-AUC Curve

In [20]:

```
model_ev = pd.DataFrame({'Model': ['Logistic Regression','NaiveBayes','VC'], 'Accuracy':
   [lr_acc_score*100,
nb_acc_score*100,grid_acc_score*100]})
model_ev
```

Out [20]:

	MODEL	ACCURACY
0	Logistic Regression	77.000000
1	Naive Bayes	76.066667
2	VC	84.800000

In [21]:

```
model_gr = pd.DataFrame({'Model': ['LR','NB','VC'], 'Accuracy': [lr_acc_score*100,
nb_acc_score*100,grid_acc_score*100]})
colors = ['blue','red','green']
plt.figure(figsize=(5,5))
plt.title(" Accuracy of three models using Voting Model")
plt.xlabel("Algorithms")
plt.ylabel("Accuracy")
plt.bar(model_gr['Model'],model_gr['Accuracy'],color = colors)
plt.show()
```

The accuracy of the three models is compared in Figure 7.3. Logistic regression shows an accuracy of 77%. Naive Bayes classifier

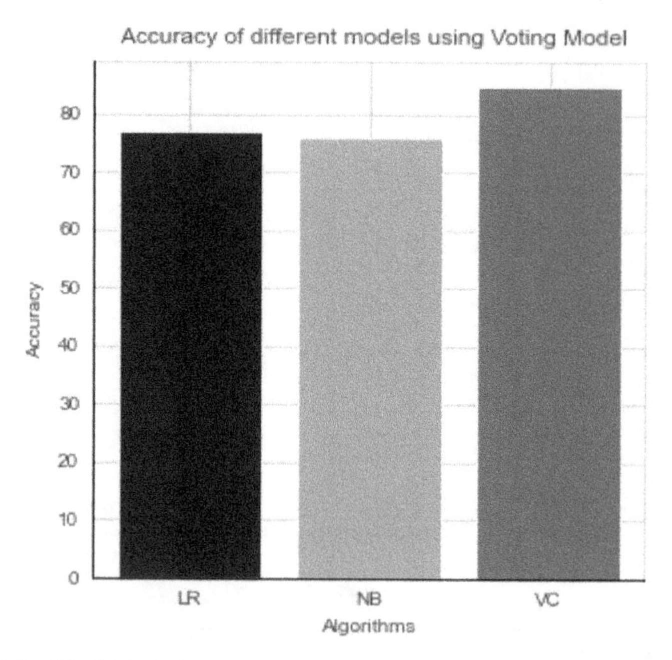

Figure 7.3 Accuracy of the Three Models

shows an accuracy of 76.07%. Voting classifier, the combination of logistic regression and Naive Bayes, shows the highest accuracy of 84.80%.

Conclusion

This work focused on predicting the employee turnover using a single voting model. Data preprocessing was done at the initial stage. The employee turnover was predicted with two different classifiers, logistic regression and Naive Bayes separately. Then the two classifiers were combined to form a single voting model to improve the performance of the model. The accuracy of the model with logistic regression and Naive Bayes is found to be 77% and 76.07%, respectively. The accuracy of the single voting model is found to be 84.80%. The experimental results show that the combination of the two classifiers outperforms when the models were run separately.

References

[1] S. Karande, and L. Shyamala, "Prediction of employee turnover using ensemble learning." In Y. C. Hu, S. Tiwari, K. Mishra, and M. Trivedi (eds), *Ambient Communications and Computer Systems. Advances in Intelligent Systems and Computing*, vol. 904. Singapore: Springer, 2019.

[2] P. Ajit, and R. Punnoose, "Prediction of employee turnover in organizations using machine learning algorithms." *Algorithms*, vol. 4, no. 5, p. C5, 2016.

[3] K. Arai et al. (Eds.), "Employee turnover prediction with machine learning: A reliable approach." IntelliSys 2018, AISC 869, pp. 737–758, 2019. Springer Nature Switzerland AG 2019.

[4] McKinley Stacker IV, IBM Watson analytics. Sample data: HR employee attrition and performance [Data file]. Retrieved from https://www.ibm.com/communities/analytics/watson-analytics-blog/hr-employee-attrition/ (2015)

[5] Xiang Gao, Junhao Wen, and Cheng Zhang, "An Improved Random Forest Algorithm for Predicting Employee Turnover." *Mathematical Problems in Engineering*, vol. 2019, Article ID 4140707, 12 pages, 2019, doi: 10.1155/2019/4140707.

[6] T. H. Feeley, J. Hwang, and G. A. Barnett, "Predicting employee turnover from friendship networks." *Journal of Applied Communication Research*, vol. 36, no. 1, pp. 56–73, 2008.

[7] P. A. Gloor, A. Fronzetti Colladon, F. Grippa, and G. Giacomelli, "Forecasting managerial turnover through e-mail based social network analysis." *Computers in Human Behavior*, vol. 71, pp. 343–352, 2017.

[8] W. C. Hong, P.F. Pai, Y. Huang, and L. Yang, "Application of support vector machines in predicting employee turnover based on job performance." In *Proceedings of the International Conference on Advances in Natural Computation*, pp. 668–674, Springer, Berlin, Germany, 2005.

[9] H.W. Kao, S.W. Lin, and S.Y. Wan, "Applying decision tree to predict nursing turnover – A case study in a public hospital." *The Journal of Taiwan Association for Medical Informatics*, vol. 21, no. 4, pp. 15–29, 2012.

[10] H. Tran, "The impact of pay satisfaction and school achievement on high school principals' turnover intentions." *Educational Management Administration and Leadership*, vol. 45, no. 4, pp. 279–290, 2016.

[11] L. J. Labrague, D. Gloe, D. M. McEnroe, K. Konstantinos, and P. Colet, "Factors influencing turnover intention among registered nurses in Samar Philippines." *Applied Nursing Research*, vol. 39, pp. 200–206, 2018.

[12] J. Rodriguez, L. Kuncheva, and C. Alonso, "Rotation forest: A new classifier ensemble method." *IEEE Transactions on Pattern Analysis and Machine Intelligence*, vol. 28, no. 10, pp. 1619–1630, 2006.

[13] A. Bagnall, M. Flynn, J. Large, J. Lines, A. Bostrom, and G. Cawley, "Is rotation forest the best classifier for problems with continuous features?" School of Computing Sciences, University of East Anglia, April 2020.

[14] Riyanto Jayadi, Hafizh M. Firmantyo, Muhammad T. J. Dzaka, Muhammad F. Suaidy, and Alfitra M. Putra, "Employee performance prediction using Naive Bayes." *International Journal of Advanced Trends in Computer Science and Engineering*, vol. 8, no. 6, pp. 3031–3035, November–December 2019.

[15] X. Cai et al., "DBGE: Employee turnover prediction based on dynamic bipartite graph embedding." *IEEE Access*, vol. 8, pp. 10390–10402, 2020, doi: 10.1109/ACCESS.2020.2965544.

8

A Novel Implementation of Sentiment Analysis Toward Data Science

VIJAYALAKSHMI SARAVANAN[1], ISHPREET SINGH[2], EMANUEL SZAREK[2], JEREON HAK[2], AND ANJU S. PILLAI[3]

[1]Faculty at Rochester Institute of Technology, Rochester, New York, USA
[2]M.S. Students, Rochester Institute of Technology, Rochester, New York, USA
[3]Amrita Vishwa Vidyapeetham, Coimbatore, India

Contents

Introduction

Sentimental analysis (SA) is used to study the emotions of the people, attitude of a person on a particular event/topic, and general behavior during conversation. SA is also known as opinion mining, which refers to text mining and natural language processing (NLP).

DOI: 10.1201/9781003119838-8

Essentially data scientists can make use of these tools to analyze the sentiment of large networks.

Communication is the essential component to progress on corporate level and personal level. There are numerous ways of communication, such as speech, hand signals, text, etc., among which speech is deemed the most significant medium for human interaction. The world is moving to digital era and has influenced the communication techniques too. Speech recognition systems, such as Amazon's Alexa, Apple's Siri, Google assistant, and many more, are ubiquitous. There may be variations in the semantics, but the central idea remains the same. The progressive advancements in technology made all these advents possible. When phone calls, messages, SMS, etc., became an integral part of human life with the invention of mobile phones, the innovation in SMS technology via speech recognition brought a tremendous change, by which voice messages are reborn as text messages. The translations such as text-to-speech (TTS) and speech-to-text (STT) have made user interaction with devices possible, assisting disabled people and encouraging users to carry out more compelling local and remote services.

STT has continued to get better as the world is rapidly advancing with technological innovations. As compared with many existing technologies, speech recognition was left to the imagination of the science fiction community. The speech recognition is deeply rooted in a research carried out by Bell Labs' researchers back in 1952; it was designed to recognize only numbers [1]. This system was discovered by Alexander Graham Bell through a process that converts sound waves into electrical impulses and is one of the novel innovations until date [2]. Since then, speech recognition has been continuing to grow and spread its reach. By 2011, easier, real-time, and faster way of communication with Apple devices was possible by the launch of Siri and to date, Amazon's Alexa and Google's Home are the most preferred voice-based virtual assistants used by the people across the globe. It has transformed the way the companies run their businesses. But, companies lack in finding a reliable means of computing the human quality of spoken conversations with their customers to help employees with the way they handle inventory. During inventory management, certain words are most likely being re-used repeatedly. These data can hold valuable information because language is very important depending on how you use it. Words carry meaning and we can distinguish how someone may be feeling depending on what words they are using.

The speech converted to text finds further more purposes such as SA. SA is a combination of methods, techniques, and tools to recognize and group opinions using NLP to decide whether the speaker's attitude toward a specific topic/product is positive, negative, or neutral [3]. It is very helpful in comprehending the mood of humans in many cases. In a long conversation, people render their opinions about the topic, products, social issues, movies, etc. The identification of the sentiment of the speaker can enhance the retrieval of the content to improve the usefulness and collecting the combined sentiment of a group of people on alike topic can help in setting up the general sentiment. Sentiment detection using text is an evolved research area and can tremendously influence the inventory by giving emphasis on the product reviews. Based on the emotional prosody studies, the tones of every human's voice can be distinguished by its intensity, pitch, loudness, speech rate, pauses, timbre, and many such parameters and whose deviations reveal different information from the speaker to the listener.

There are wide varieties of SA applications for social goods, such as fighting for the pandemic, preparing people for future work, promoting inclusivity, and protecting the environment. In this work, we approached SA for social goods in a different way such that one can utilize our analytic results in assistive technologies for students with learning difficulties, and person with disabled arms. We use STT tools to describe the efficiency of how our data science algorithm outperforms with our real-time application domain, inventory management systems. Further, this SA algorithm can be implemented effectively in assistive devices.

The proposed work implements STT conversion with SA to detect the condition of machine/module in an industry for machinery maintenance. Consider a case in which many laborers work in an industry where different mechanical parts are fabricated and there is a need to guarantee that every module is in working condition and in stock. To incorporate an efficient method for finding the working condition of the module, STT technique can be employed in place. By having an STT application, the details about the parts could be sent directly to a text document. We then used the obtained document to differentiate between the parts. Adding SA to this case was found to be interesting and fetching accurate results. The outcome of the SA will help the workers to be aware of parts needed to be considered on a concrete level, rather than simple guesses and predictions, thereby improving the production and down time.

The paper is organized as follows: the section "Related Work" presents a detailed literature summary about the related work done in the field, followed by background of SA in the section "Background of Sentiment Analysis". The section "Proposed Methodology" explains the methodology adopted to implement STT and SA algorithms. Results are presented in the section "Results" and the future application prospects of STT and sentiment algorithm are discussed in the section "Future Proposals of Speech-to-Text Sentiment Analysis". The paper is concluded in the section "Conclusions and Future Directions" along with the future direction of work.

Related Work

The vast literature shows that there is a lot of interest shown by the researchers on SA. SA identifies the sentiment conveyed in a text and upon analysis finds out whether the text expresses positive, negative, or neutral sentiment. Kralj Novak et al. [4] conducted a Twitter message SA based on emoji use. All messages were labeled with either a positive, neutral, or negative label. If emojis would occur several times, all occurrences were counted. They found that the use of popular emojis could be used to classify a twitter message on sentiment. A comparison was also made for twitter messages in different languages and in all languages, there were emojis that can be used to classify a message as positive, neutral, or negative.

Besides STT, an algorithm could be trained to recognize unique voices and unique words. As Saurav Bhandari et al. [5] proposed, a speech detection system could be used for unlocking and starting cars. The algorithm would recognize the specific voice of the user and recognize the designated word (or sentence) for unlocking and starting the car. The idea was to increase the security and safety against theft. To enhance safety, the proposed system could be expanded with face recognition sensor and software. The entire system could work on a Raspberry Pi. Also, there are many SA implementations using algorithms such as Naïve Base, support vector machine, decision tree, maximum entropy, and many more in the literature [6–8]. Mostafa Shaikh et al. [9] proposed a method to classify the input sentences as subjective or objective from a given document. Classical machine learning (ML) techniques are applied for the subjective parts and objective parts are discarded. A multi-modal approach to identify the sentiment of products based on audio and text was proposed by Abburi et al. [10]. A real-time speech emotion recognition based on deep neural network (DNN) to identify emotions from a one-second raw speech spectrograms was presented by Fayek et al. [11]. SA applied to political debates was considered in a research by Salah et al. [12]. A maximum entropy modeling approach to identify sentiment using natural audio streams from YouTube was presented by Kaushik et al. [13].

STT has been around for quite some time and has had many different applications. One primary use is of educational purpose, that is to make teaching easier [14–18]. In the work by Shadiev et al. [19], different ways of STT translations that helped the educational field are presented. Clearly, there are many ways in which STT can be used to enhance the learning capabilities of educators. Due to the nature of online classrooms, some teachers may feel disconnected with students. It may be hard to process what the students are feeling. A possible future study could be carried out to use a combination of the STT with SA to see if there can be different techniques applied to help the students learn better.

Social media has become a sort of Wild West in our society. There are not a lot of moderators or controls in place. For most cases, people may say or express whatever they want. This creates a massive forum of both positive and negative opinions, which are ripe with data. This should be very helpful for those attempting to test the waters of a new product. An article that did SA to evaluate social media posts using YouTube gamers, such as PewDiePie, found some interesting results. Whenever content was posted, usually some sort of repost from YouTube, on Facebook, it was generally not very well received and most of the comments were negative [20]. Market analysts can get a better picture of what brings people together and use it to their advantage to create advertisements based on what brought on these happy comments.

The application for SA is endless. Ben-Porat et al. have conducted research to predict action from free text [21]. A collection of text including text messages, Facebook posts, tweets, etc. is analyzed as an input into the algorithm to predict an action made in a game [21]. The study proved successful and there was a correlation between free text and the actions a person would take in a game. The results were biased toward certain genders and the data analyzed were primarily college-aged students but still opened the possibility with a larger dataset and a robust ML algorithm that can successfully analyze and predict a user's behavior. This can be a very useful tool for marketers along with advertisers to place the proper advertisements for users.

Going beyond sentiment, we can also look at the data collected from STT algorithms. Research done by Gillick and Bamman et al. predicted the lyrics for a given audio input [22,23]. The project takes an audio source and removes any speech leaving background sound. From the background audio input into the algorithm we can create musical lyrics based on the tones. Data produced from the STT application can be used to train and create lyrics. This can assist musicians and vocalists writing lyrics for their music. Perhaps one day we'll see producers creating beats and an algorithm developing lyrics to become the next biggest virtual artist.

Background of Sentiment Analysis

Sentimental analysis identifies the sentiment in the expressed text and then it analyzes the opinion, which is shortly referred as SA. Speech recognition has a lot of different factors. Based on the uttered words, words are classified as isolated word, connected word, continuous speech, and spontaneous speech. Isolated word is what it sounds like, when a single word is spoken. There will be silence, a word will be spoken, and then silence again. Connected word is having two words instead of one. One would utter a word and then a second word right after it with minimal pause. Continuous speech is described as talking naturally and these are considered as one of the difficult tasks because identification of utterance boundaries is done through unique sound and special methods [24]. Finally, spontaneous speech is impulsive speech that you would think of on the spot and just start yapping away. This feature is good at detecting when the user is stringing different words together and when he/she is not.

Depending on the place we grew up, everyone's speech dialect is different. Sometimes, for humans it can be challenging to understand each other. This becomes more challenging for the speech software to decode. Humans have such distinct voices and dialects that make it harder for the software to understand what is being said. It is due to these differences that there are two types of speaker models. (1) There are speaker independent models, which are built for the masses, so that they are able to understand a large array of speakers effectively. But is hard to implement and is costly, and it renders reduced accuracy in comparison with speaker-dependent systems. One of the factors for its wide usage is its flexibility. (2) Speaker-dependent systems are the opposite of the independent systems and are built specifically for one target individual. These systems lack flexibility but are preferred as they provide easy development and more accuracy, and are cheaper. For single speaker model, it exhibits high accuracy, while for multi-speaker model, the speaker-dependent system produces less accuracy [24].

When it comes to opinion, people usually have a stance on whether they agree or support something, or whether they disagree and are against something. It seems like the more the opinion is supported, the stronger the support/hate is. With SA,it tries to differentiate between the two and put them into a positive review or negative review. This can be done in two different ways: direct or comparison opinions. With direct opinion, we get the view of the product as positive or negative directly. For example, "Life style of people at rural area is poor" expresses a direct opinion. Meanwhile, for comparison, we compare the subject with other objects of similar class. For example, "The life style of people in *village-A* is better than that of people in *village-B*" expresses a comparison.

In Mostafa et al.'s research [9], authors presented the sentence in each document labeled with only subjective part and discarded objective part. After that, they applied the classical ML algorithms on subjective parts, which is time consuming at the sentence level and not an easy job to test them. To perform STT on SA, we have used the following methods: – logistic regression and comparison are made to find the efficiency of the proposed algorithm.

Proposed Methodology

Speech to Text Model

STT model is built on the concept of an on-line STT engine [25]. The primary objective of any speech translation system is to make it possible for a machine to listen, recognize, and produce actions based on the spoken words. The user voice at run-time is captured via a microphone to process the sampled speech. This allows recognizing the uttered word by the user. The identified word is then conerted into output as a text. For large volume of input speech signal, the identified words can be stored in a file for further analysis and use. Large and complex systems can adopt different choices for data entry. The STT algorithm was built around the Google STT Python library. The Google API allows us to keep the algorithm very lightweight and scalable. The Google API does STT efficiently allowing the developers to focus on adding additional features to the algorithm and making the product user friendly, as shown in Figure 8.1. In the proposed work, the STT function in Python is used and the users microphone is used as input.

Additional features such as SA and widgets were added to allow users to use the algorithm with ease and allow engineers to look at the data and respond accurately. The STT and SA algorithm is a user-friendly codebase that can be implemented across multiple devices.

There are some negative impacts of the proposed algorithm, since it relies on Google STT API. If Google had any outages or downtime, the STT would cease to function. Since STT relies on an API, the ability to customize the library and improvements are limited for future development. STT algorithm is very sensitive and if there is loud background noise, the model will exhibit reduced performance by producing inaccurate texts from the processed sound. SA has its own set of negatives with a small training set that was custom built for engineers. Due to the small training set, the error rate contains

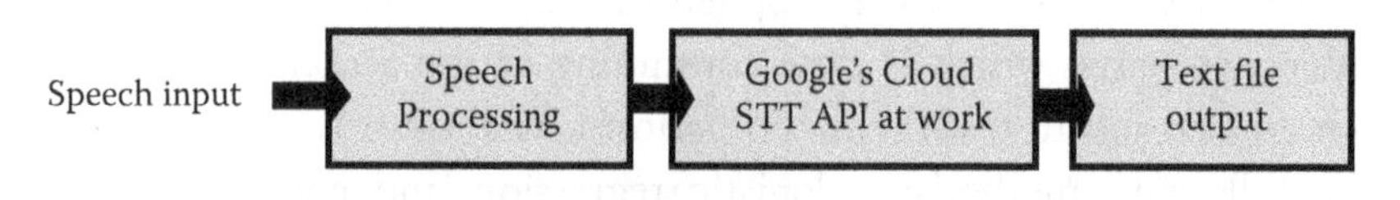

Figure 8.1 General Representation of the STT Model

biases and has an error rate. Overtime data collected can be used to improve the SA algorithm. The SA algorithm can also be improved to a more robust algorithm with a higher accuracy rate. Other data mining features can be added in the future.

Dataset: https://www.kaggle.com/c/tensorflow-speech-recognition-challenge
Methods: The software used is Python. Specifically, the libraries libROSA and SciPyis are used to process the audio signals.

Implementation of Sentiment Analysis

SA is also referred as opinion mining, as it combines volumes of opinion on a topic/product. Text mining is a technique used for opinion mining. The various processes involved in SA are shown in Figure 8.2. The SA algorithm takes input from online user and processes the raw data to extract the features from the input signal. From the extracted features, adjectives are identified to detect the emotions of the user involved in the conversation.

In the proposed work, SA is implemented using a logistic regression algorithm. Logistic regression is a classification model and is one of the dominant algorithms for classification used in industry. The steps of logic regression implementation are shown in Figure 8.3.

The input to the algorithm is maintenance data set from a machinery maintenance industry. This maintenance data are the input vector x. Vector of weights w are created to adjust accordingly to predict the output. Through the adjusting of weights, the algorithm learns to classify the output correctly. The net input is referred as the dot product of input vector and weight. Sigmoid function is used to fix the net input in a range of [0,1], for classification of the input as either positive or negative. Defining the threshold function helps in accurate classification of the output.

Architecture of Sentiment Analysis

Architecture of SA exhibits the steps involved in concluding reviews from the user input. Figure 8.4 shows the detailed steps of architecture of SA.

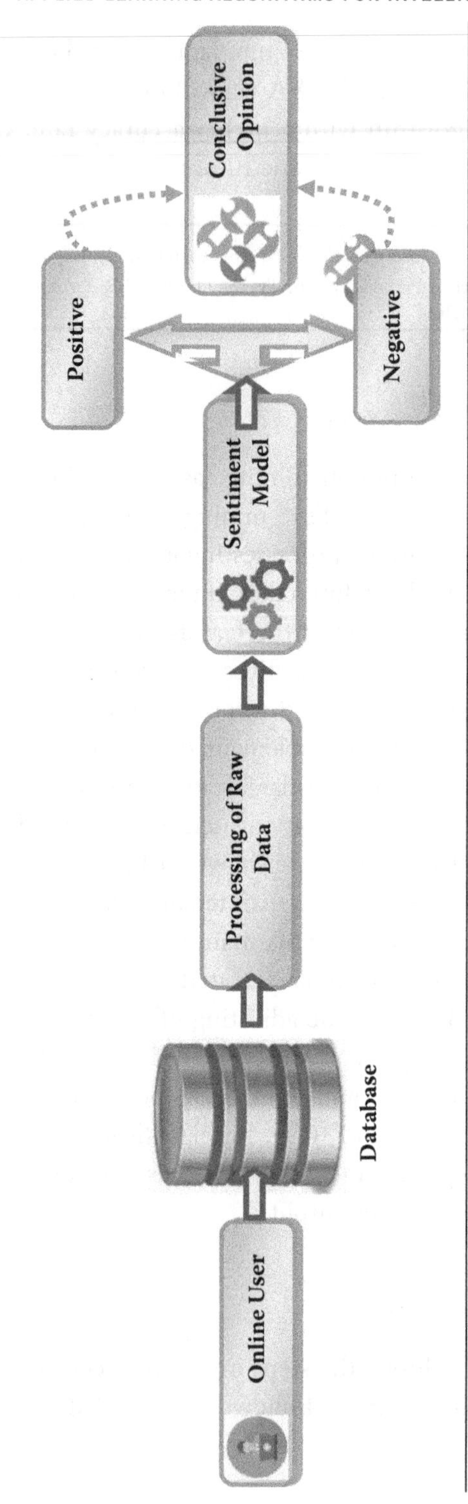

Figure 8.2 General Process of Implementation of SA

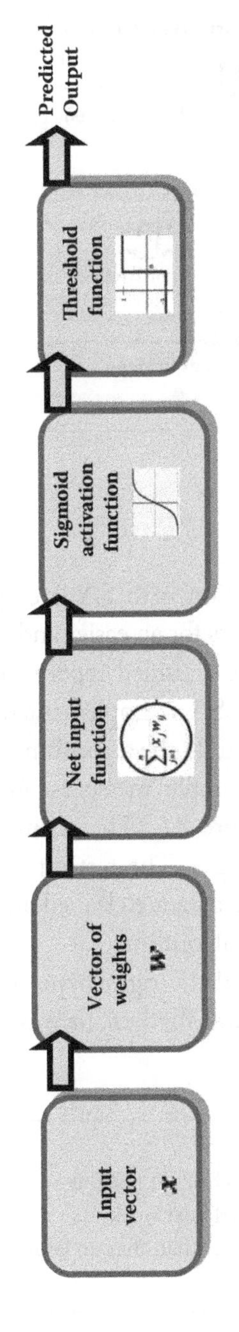

Figure 8.3 Steps of Implementation of Logic Regression Algorithm

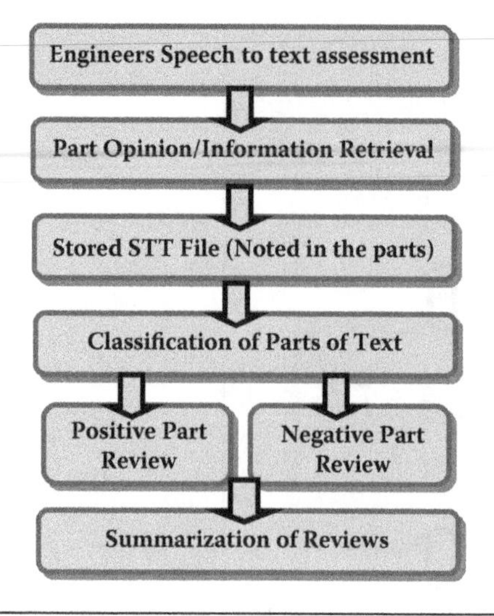

Figure 8.4 Architecture of SA

Results

In this work, the STT translation with SA is employed to report the module/part failure by mechanics for an easier and accurate classification of their findings. The negative classified reports could be reviewed by engineers to identify patterns. SA is implemented using a logistic regression algorithm. In order to train the algorithm, a training set and test file are created containing example messages about maintenance, as shown in Figure 8.5 and Figure 8.6. The training set consists of 75 positive and 75 negative messages. The test set consists of 50 positive and 50 negative messages. The data are cleaned and vectorized in order to train the logistic regression algorithm.

An important step is to find the right hyper parameter *C*. Several values for *C* have been tried and the best fit was chosen, as shown in Table 8.1.

Part 1 appears to have no issues
Part 2 seems like it has a few scratches but nothing worth removing
Part 3 is very sharp and can be used next week
Part 4 needs to be buffered but other than that no issues
Part 5 looks new

Figure 8.5 Snapshot of Positive Training Set

Part 152 crappy ridges and needs new bearings
Part 153 makes me question who approved thiscrap
Part 154 was just morejunk
Part 155 feels like it was put together in seconds and feels super cheap

Figure 8.6 Snapshot of Negative Training Set

Table 8.1 Accuracy for Different C Values

C VALUE	ACCURACY
0.01	0.76
0.05	0.80
0.25	0.82
0.50	0.84
1.0	0.84
1.5	0.86
2.0	0.86
3.0	0.86

All values higher than $C = 1.5$ do not add to more accuracy. The value of $C = 1.5$ is used to train the algorithm. After training the algorithm, it was tested and evaluated based on the accuracy. The accuracy of the algorithm is shown in Figure 8.7. The accuracy of the proposed model is found to be 81% and can be considered as sufficient for practical use.

Upon training and testing, the logistic regression algorithm is integrated with the STT translation algorithm. This algorithm uses the STT function in Python and uses the user's microphone as input. The output is both written to a .txt file and put through the SA. The .txt output can be used to fill in the mechanics' work order and the SA can be sent to the engineering department. Other uses may be applicable too. To make the STT algorithm even more accessible, it

Final Accuracy: 0.81

Figure 8.7 Final Accuracy

['part is looking good', 'no problems here', 'doors function very well']

Figure 8.8 Final Model Input Example

was supplied with a button to record the user's microphone for a set amount of time. Using the slider, the amount of time can be adjusted. In the figure, the slider was set to 40 seconds.

The STT algorithm was used to record multiple sentences and the output was put through the SA. In Figure 8.8, a few example lines are shown.

The algorithm classified all lines as positive as it should. A message gets classified as positive by a "1" and negative by a ""0." Multiple tests were conducted and the algorithm seems to classify the microphone most of the time correctly.

In our initial experimentation and testing, we were very successful in running the STT algorithm. The STT itself is powered by the Google STT API. The API is very powerful and accurate in taking the speech input and giving a rapid output of text. By combining the STT with our trained SA, we were able to get an accuracy rate of 81%. The output of the SA will help provide mechanics the most critical situations to work on.

Future Proposals of Speech-to-Text Sentiment Analysis

The application of the STT SA is proposed for mechanics to report their findings while maintaining machinery. The STT algorithm makes it easier for mechanics to report the condition of the machines rather than typing it on a keyboard. It is useful when they are not close to a computer or even when the computer is near, as their hands could be dirty due to maintenance or repairing of the machinery. The SA can then be used by engineers who oversee maintenance. By focusing on the analysis, the machine conditions can be classified as negative or positive and it becomes easier to know the status of the machinery. This tool can go hand in hand with real-time monitoring tools that monitor the status of machinery through the sensor data. Combining the SA of findings and the real-time monitoring tools could provide a good understanding of the status of the machinery and will improve the production.

Other than the maintenance, there are more applications for the STT SA software. When you pay at a restaurant by card, the card reader could perhaps ask your opinion and record it. The file is sent to a computer, which runs the speech file through the software and puts both the text and the SA in a database. Some people might feel embarrassed to give their honest opinion when the waitress is standing beside them. The waitress leaving the card reader on the table and coming back after the customer has paid could fix this.

The same could be applied when rating an app or an Uber driver. The rating with stars or numbers should not necessarily have to be operated by speech. However, when there is a comment box, this could be replaced by speech. The sentiment of the message can be used to rate the app or the Uber driver as an addition to the rating by stars or numbers.

Conclusion and Future Directions

This works presents the generalized model of STT on SA. In this proposed work, we have implemented a simple system called inventory management systems to perform the above-mentioned tasks to show how this can be utilized for social goods, such as assistive devices. This system works well with smaller datasets. We are currently working on a larger dataset and increasing scalability of the proposed system.

The goal of the proposed work is to assist engineers to rapidly record the system status and equipping sentiment analysis (SA) to assess the status of the mechanical modules/parts. This is achieved by implementing a speech-to-text (STT) translation algorithm to output the corresponding texts recorded by the mechanic. The STT algorithm is integrated with SA algorithm implemented using logic regression classifier to group the output as either positive or negative review. With the assessment provided by SA, mechanics can engage the negative observances and fix the issues. The proposed work can also assist with analytics and record the number of negative situations engineers encounter each day.

This work is an amazing asset for engineers to use. The software developed is very lightweight and user friendly providing a high rate of accuracy. We can continue improving the algorithms by increasing the rate of accuracy of the observances. As the input data grow, they can be collected and added to the training set to improve our algorithm.

Another improvement that can be made is looking at a more robust SA algorithm, such as a neural network to provide a higher accuracy rate. Creating a more robust algorithm will be beneficial for mechanics to handle critical situations. We can also include analytics and data mining functionality with more data to improve the accuracy of producing SA output. Data mining at minimum would include information about the parts that are continuously failing and being replaced or the number of issues mechanics are dealing with on a daily basis. Another improvement is adding multi-device support for the proposed work with the possibility of rolling out mobile-ready applications that mechanics can use.

References

[1] Kikel, Chris. (2019). "A Brief History of Voice Recognition Technology," Total Voice Technologies. https://www.totalvoicetech.com/a-brief-history-of-voice-recognition-technology/.

[2] Ghai, Wiqas, and Singh, Navdeep. (2012). "Literature Review on Automatic Speech Recognition," *International Journal of Computer Applications*, vol. 41, no. 8, pp. 42–50. https://doi.org/10.5120/5565-7646.

[3] Mäntylä, Mika V., Graziotin, Daniel, and Kuutila, Miikka. (2018). "The Evolution of Sentiment Analysis – A Review of Research Topics, Venues, and Top Cited Papers," *Computer Science Review*, vol. 27, pp. 16–32. https://doi.org/https://doi.org/10.1016/j.cosrev.2017.10.002.

[4] Kralj, Novak P., Smailović, J., Sluban, B., and Mozetič I. (2015). "Sentiment of Emojis," *PLoS ONE*, vol. 10, no. 12, e0144296. doi: 10.1371/journal.pone.0144296. https://journals.plos.org/plosone/article?id=10.1371/journal.pone.0144296.

[5] Bhandari, Saurav, Jain, Saiyam, Jambhale, Sagar, and Khapare, Ajinkya. (2017). "Voice Recognition System for Automobile Safety." *International Research Journal of Engineering and Technology*, vol. 4, no. 2, https://www.irjet.net/archives/V4/i2/IRJET-V4I2303.pdf.

[6] Pang, B., and Lee, L. (2014). "A Sentimental Education: Sentiment Analysis Using Subjectivity Summarization Based on Minimum Cuts," In *Proceedings of the 42nd Annual Meeting on Association for Computational Linguistics*, p. 271, Association for Computational Linguistics.

[7] Pang, B., and Lee, L. (2005). "Seeing Stars: Exploiting Class Relationships for Sentiment Categorization with Respect to Rating Scales," In *Proceedings of the 43rd Annual Meeting on Association for Computational Linguistics*, pp. 115–124, Association for Computational Linguistics.

[8] Pang, B., Lee, L., and Vaithyanathan, S. (2002). "Thumbs Up?: Sentiment Classification Using Machine Learning Techniques," In *Proceedings of the ACL-02 Conference on Empirical Methods in Natural Language Processing*, vol. 10, pp. 79–86, Association for Computational Linguistics.

[9] Shaikh, M., Prendinger, H., and Mitsuru, I. (2007). "Assessing Sentiment of Text by Semantic Dependency and Contextual Valence Analysis," *Affective Computing and Intelligent Interaction*, ACII 2007, LNCS 4738, pp. 191–202.

[10] Abburi, Harika, Gangashetty, Suryakanth V., Shrivastava, Manish, and Mamidi, Radhika. (2017). "Audio and Text Based Multimodal Sentiment Analysis Using Features Extracted from Selective Regions and Deep Neural Networks," MS Thesis. International Institute of Information Technology, Hyderabad, India.

[11] Fayek, H. M., Lech, M., and Cavedon, L. (2015). "Towards Real-Time Speech Emotion Recognition Using Deep Neural Networks," *2015 9th International Conference on Signal Processing and Communication Systems (ICSPCS)*, Cairns, QLD, pp. 1–5, doi: 10.1109/ICSPCS.2015.7391796.

[12] Salah, Saleh, and Ibrahim, Zaher. (2014). "Machine Learning and Sentiment Analysis Approaches for the Analysis of Parliamentary Debates," Thesis.

[13] Kaushik, L., Sangwan, A., and Hansen, J. H. L. (2013). "Sentiment Extraction from Natural Audio Streams," *2013 IEEE International Conference on Acoustics, Speech and Signal Processing*, Vancouver, BC, pp. 8485–8489, doi: 10.1109/ICASSP.2013.6639321.

[14] Nisbet, P., and Wilson, A. (2002). "Introducing Speech Recognition in Schools," Edinburgh, UK: CALL Centre, University of Edinburgh.

[15] Nisbet, P., Wilson, A., and Aitken, S. (2005). "Speech recognition for students with disabilities," Proceedings of the Inclusive and Supportive Education Congress, ISEC 2005 Conference, Delph, UK: Inclusive Technology.

[16] Nisbet, P., Wilson, A., and Balfour, F. (2008). *Introducing Speech Recognition in Schools: Using Dragon Naturally Speaking*, Edinburgh, UK: CALL Centre, University of Edinburgh.

[17] Petta, T. D., and Woloshyn, V. E. (2001). "Voice Recognition for On-line Literacy: Continuous Voice Recognition Technology in Adult Literacy Training," *Education and Information Technologies*, vol. 6, no. 4, pp. 225–240.

[18] Ranchal, R., Taber-Doughty, T., Guo, Y., Bain, K., Martin, H., Robinson, J., and Duerstock, B. (2013). "Using Speech Recognition for Real-Time Captioning and Lecture transcription in the classroom," *IEEE Transactions on Learning Technologies*, vol. 6, no. 4, pp. 299–311.

[19] Shadiev, Rustam, Hwang, Wu-Yuin, Chen, Nian-Shing, and Huang, Yueh-Min. (2014). "Review of Speech-to-Text Recognition Technology for Enhancing Learning," *Journal of Educational Technology & Society*, vol. 17, no. 4, pp. 65–84. Accessed April 8, 2020. www.jstor.org/stable/jeductechsoci.17.4.65.

[20] Poecze, Flora, Claus Ebster, and Christine Strauss. (2018). "Social Media Metrics and Sentiment Analysis to Evaluate the Effectiveness of Social Media Posts," *Procedia Computer Science*, vol. 130, pp. 660–666. https://doi.org/10.1016/j.procs.2018.04.117.

[21] Ben-Porat, Omar, and Sharon Hirsch. (2020). "Predicting Strategic Behavior from Free Text," *Journal of Artificial Intelligence Research*. https://arxiv.org/pdf/2004.02973.pdf.

[22] Gillick, Jon, and David Bamman. (2019). "Breaking Speech Recognizers to Imagine Lyrics," *NeurIPS 2019, Workshop on Machine Learning for Creativity and Design*, December 15. https://arxiv.org/pdf/1912.06979.pdf.

[23] Kaushik, Abhishek, Anchal Kaushik, and Sudhanshu Naithani. (2015). "A Study on Sentiment Analysis: Methods and Tool," *International Journal of Science and Research*, vol. 4, no. 12, pp. 287–292. https://pdfs.semanticscholar.org/c151/dfad8c1bf88b0afc716758c77d533ded7dd0.pdf.

[24] Khilari, Prachi, and Bhope, V. P. (2015). "A Review on Speech to Text Conversion Methods," *International Journal of Advanced Research in Computer Engineering and Technology*, vol. 4, no. 7, pp. 3067–3072. http://ijarcet.org/wp-content/uploads/IJARCET-VOL-4-ISSUE-7-3067-3072.pdf.

[25] B. Raghavendhar Reddy, E. Mahender. (2013). "Speech to Text Conversion Using Android Platform", *International Journal of Engineering Research and Applications*, ISSN: 2248-9622, vol. 3, no. 1.

9

CONSPECTUS OF K-MEANS CLUSTERING ALGORITHM

USHA SAKTHIVEL[1], A. P. JYOTHI[2], N. SUSILA[3], AND T. SHEELA[4]

[1]*Dean, Dept. of Research and Innovation, Rajeswari College of Engineering, Bengaluru, India & Prof. & Head, Dept. of Computer Science & Engineering, Visvesvaraya Technological University, Belagavi, Karnataka, India*

[2]*Associate Professor, Department of CSE, Visvesvaraya Technological University Belagavi, Karnataka, India*

[3]*Prof. & Head, Department of IT, SKCET, Coimbatore, India*

[4]*Prof. & Head, Department of IT, SSEC, Chennai, India*

Contents

DOI: 10.1201/9781003119838-9

Introduction

The expression "K-means" was originally utilized by James MacQueen in 1967 as a major aspect of one of the manuscripts. The standard calculation was likewise utilized in Bell Labs as a feature method in beat program tweak in 1957. It was likewise distributed by E. W. Forgy.

The majority of unsubstantiated erudition-based purposes use the sub-domain as clustering. It is the way toward gathering information tests together into clusters dependent on a specific element that they share – precisely the motivation behind unaided learning in any case. We are given an informational index of things, with specific highlights, and qualities of these highlights. The undertaking is to classify those things into K gatherings. To accomplish this, we will utilize the K-means calculation, an unsupervised learning algorithm. To compute that similitude, we will utilize the Euclidean separation as estimation.

Background Study

Tajunisha et al. [1]. Execution analysis of K-means with distinctive instatement techniques for high-dimensional information utilizes principal component analysis (PCA) to measure decrease and discover starting bunch communities. The variable with the most elevated Eigen esteem is determined utilizing PCA and is taken as first head segment along which apportioning is done, on the premise of which K subsets are shaped and K middle qualities are taken as beginning K communities.

Bouhmala et al. [2]. Joined genetic algorithm (GA) and K-means (GAKM) are used to improve the nature of bunches framed and accelerate their hunt procedure. The presentation of GAKM is tried over the datasets, for example, iris, glass, and so on, which have been taken from machine learning vault. The trial results have demonstrated that GAKM joins quicker while contrasting with standard GA. Despite the fact that this calculation neglected to catch the best nature of bunches, it is unsatisfactory for expanding both homogeneity and heterogeneity inside same bunches and with various groups individually.

Rouhollah et al. [3]. Anticipated a model called GA grouping for improving K-means calculation. The calculation has been performed on notable datasets to be specific iris, raw petroleum. The test results give bunching standard μ; the lower estimation of μ gives better bunch arrangement of information contrasted with customary K-means bunching calculation.

Jyothi A. P. et al. [4]. The paper gives an inside and out examination of the current grouping methods alongside their adequacy in vitality proficiency.

N. Kaur et al. [5]. This paper considered up gradation of the customary K-means by presenting ranking technique. The creator presents ranking technique to conquer the inadequacy of more execution time taken by customary K-means. The ranking method is an approach to discover the event of comparative information and to improve search viability. The device used to execute the improved calculation is Visual Studio 2008 utilizing C#. The focal points of K-means are likewise investigated in this paper. The creator discovers K-means as quick, powerful, and simple justifiable calculation. Also, the groups are non-various leveled in nature and are not covering in nature. The procedure utilized in the calculation takes understudy marks as informational index and afterward, the starting centroid is chosen. Euclidean separation is then determined from centroid for every information object. At that point, the edge esteem is set for every informational collection. Positioning method is applied straightaway and at last, the groups are made dependent on least separation between the information point and the centroid.

Sandeep Rana et al. [6]. This paper proposed another hybrid sequential grouping approach. They have utilized particle swarm optimization (PSO) and K-means calculation in grouping for information bunching. This methodology was proposed to beat the disadvantages of the two calculations, as well as to improve the grouping and abstain from being deteriorated. Four sorts of informational indexes have been tried so as to get relative outcomes. For examination reason, extraordinary calculations, for example, PSO, K-means, hybrid K-means PSO, what's more, hybrid K-means + GA were thought of. The proposed calculation produces progressively precise and vigorous bunching results.

Jyothi A. P. et al. [7]. The paper presents a novel logical demonstration of CFCLP that uses information conglomeration by a novel combinatorial methodology utilizing straight programming for developing up with another bunching instrument. An advanced scientific demonstration is utilized for planning and its calculation being actualized on MEMSIC hubs to find that CFCLP is equipped for keeping up great harmony between vitality sparing execution and information conveyance among the sensor hubs for longer cycles.

Bara'a Ali Attea et al. [8]. The paper found that exhibition of bunching calculations debases with an ever-increasing number of covers among bunches in an informational index. These realities have inspired to build up a Fluffy Multi-Target Molecule Swarm Streamlining System (FMOPSO) in an inventive style for information bunching, which can convey more viable outcomes than cutting-edge grouping calculations. To find out the predominance of the proposed calculation, a number of measurable tests have been completed on an assortment of numerical and all out genuine informational indexes.

Jyothi A. P. et al. [9]. The paper presents an epic multi-stage enhancement demonstration for tackling huge scope vitality advancement issue and along these lines, defined conditional ideal bunching setup is examined.

Chetna Sethi et al. [10]. The paper proposed a linear PCA-based crossover K-means bunching and PSO calculation PCA-K-PSO. In PCA-K-PSO calculation, the quick intermingling of K-means calculation and the worldwide look through the capacity of PSO are consolidated for bunching enormous information sets utilizing linear PCA. Better grouping outcomes can be acquired with PCA-K-PSO when contrasted with common PSO. This was viably evolved to make its utilization for productive bunching of high-dimensional informational collections.

Rui Xu et al. [11]. "Review of Clustering Algorithms." Data examination assumes a key job for understanding different marvels. Pack examination, unrefined investigation with little or on the other hand no earlier information, comprises examination created over a wide assortment of networks. The assorted variety, on one hand, furnishes us with various instruments. Then again, the abundance of alternatives creates disturbance, an audit gathering calculations for informational records appearing in measurements, software designing, and artificial intelligence (AI), and layouts their applications in a couple of benchmark informational lists, the voyaging sales rep issue, and bioinformatics, another field drawing in concentrated endeavors. A couple of solidly related focuses, closeness measure, and pack approval are additionally examined.

Jyothi A. P. et al. [12]. The paper presents a novel procedure called EPCA. The strategy gives two-level group streamlining by giving numerous credits to exact determination of cluster head. EPCA gives less unpredictable activity, quicker reaction, and better sustenance toward longer recreation adjusts.

Xiaohui et al. [13]. The paper utilized PSO for archive bunching. K-means calculation is most regularly utilized apportioning calculation for bunching enormous datasets, yet it produces neighborhood ideal arrangement. Rather than confined looking through property of K-means, PSO performs globalized search utilizing whole arrangement space. Creators utilized PSO, K-means, and half breed PSO bunching calculation on four report datasets, which are acquired from text retrieval.

Jyothi A. P. et al. [14]. The proposed framework in this paper presents one kind of topology control component utilizing a novel idea of interstellar direction toward upgrading the grouping execution in WSN. Receiving an expository exploration philosophy, the proposed framework presents two interstellar-based topology control framework, which focuses on the most extreme sparing of asset utilization of the group head. The reproduced result of the examination shows that the proposed geography control framework offers huge vitality protection execution in contrast with the current various leveled grouping plan in WSN.

Osama Abu Abbas et al. [15]. The paper shows comparison of K-means clustering algorithm with other unsupervised algorithms, namely, HC, SOM, and EM clustering algorithms.

Overview of K-Means Clustering

K-means clustering belongs to a category of unsupervised learning, a sort of unaided realizing, which is used when you have unlabeled data. The target is to calculate the bunches in the records, to get a count of get-togethers indicated by the variable K. The methodology of K-means clustering is depicted in Figure 9.1.

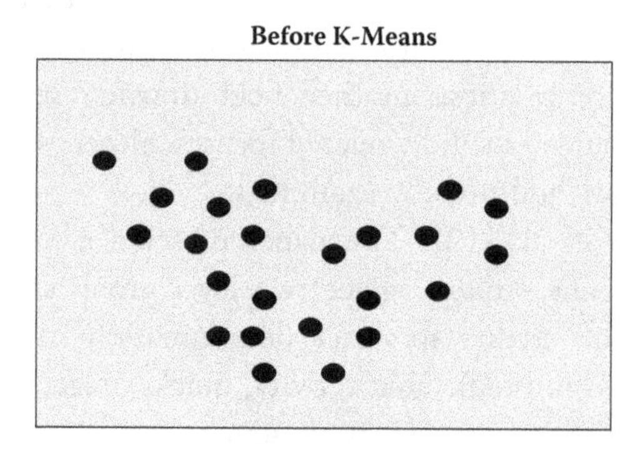

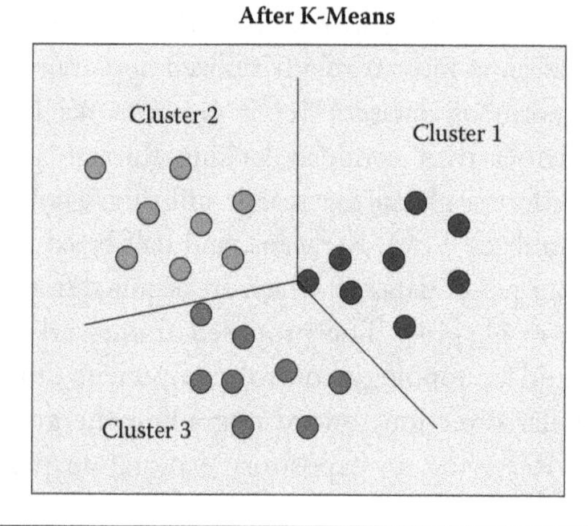

Figure 9.1 K-Means Clustering Methodology

An Insight about the Operational Working of K-Means Clustering

The scheme follows a fundamental and covenant methodology, portraying the given enlightening file by a meticulous figure of bundles predetermined previously. The indispensable proposal portrays K centroids, individual in every bundle. These ought to be set in an astuteness way because different territories cause different results. Thus, the superior choice bundles them normally far off from one another.

The accompanying phase takes each immediate have a position toward a known enlightening record and accomplice it to the nearest centroid. Exactly when no aim is forthcoming, the basic advance and an early on groupware are carried out.

As of now, we need to re-figure K new centroids (KNCs) as barycenter of the gatherings approaching, because of the point of reference advancement. After refiguring these KNCs, another blend must be prepared in the midst of comparative instructive file centres and the nearest NC.

An encircle is made. In view of this, we observe that the K centroids transform their regions slightly at a time in anticipation of nil changes. Finally, centroids do not shift to any further extent.

Steps Involved in K-Means Clustering Algorithm

Following are the steps involved:

- K centroids are formed arbitrarily.
- K-Means allocate each information point in the information set so that they are near the centroid.
- Then K-means reconfigure the centroids by averaging out the information and pronouncing the fresh centroid with respect to all information points assigned to that centroid's cluster, resulting in reduction of the overall intra-cluster dissent compared to the preceding step.
- Rehash Steps 2 and 3 until the centroids don't shift to any further extent.

Real-Time Project Implementation on Synchronic Health Fitness Detector of Combatants in Armed Conflicts Using K-Means Clustering Algorithm

The aim of this project is to track the location of combatants in armed conflicts and to provide synchronic health fitness monitoring using K-means clustering algorithm.

Proposed Work

The aim of the proposed work is to form clusters. In this project called troops are formed using K-means clustering algorithm. The chief of armed forces acts as cluster head, who responsible for data aggregation from all combatants in that troop. Use of K-means clustering algorithm plays a vital role in this project and acts as a very important proposal in providing aid to the army personnel in case of dread and emergency situations.

The project results help in locating the combatant and checking and updating his fitness parameters to chief of armed forces and onto the army control room. Chief (armed forces) of one troop can communicate with chief (armed forces) of another troop by using the NRF module. Combatants in one troop can communicate and also provide aid in case of any crisis.

The collected data by chief of armed forces will be uploaded on the cloud to army control room to provide necessary aid to the combatant.

Figure 9.2 shows the communication model of the proposed work consisting of combatant segment, chief of armed forces segment and army control room.

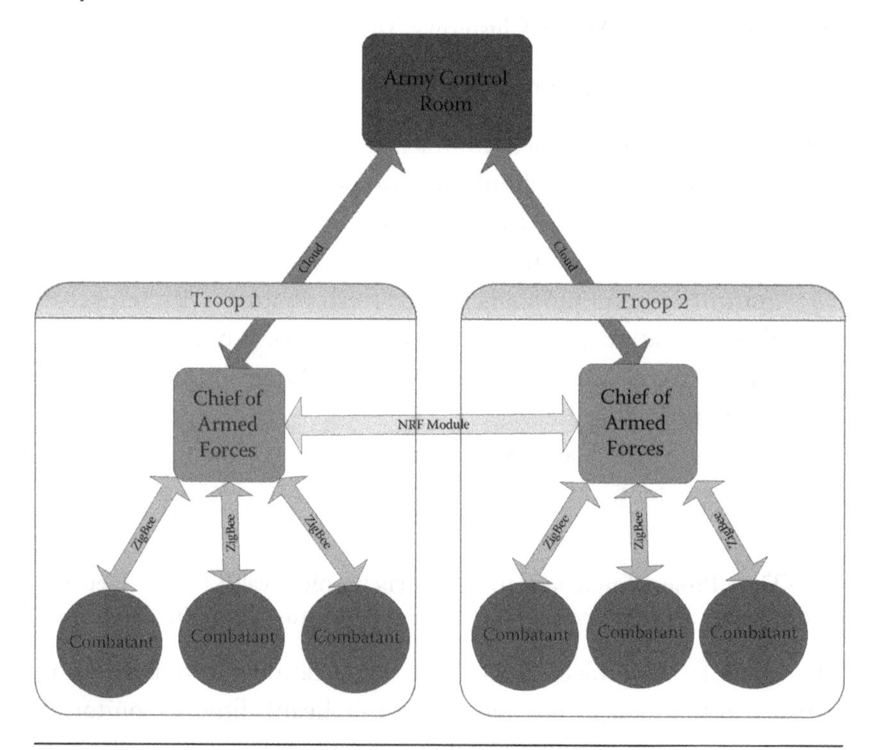

Figure 9.2 Communication Model of the Proposed Work

Each segment is discussed in the "Design Methodology and Implementation" section.

Design Methodology and Implementation

Combatant's Segment This segment has a suitable sensory unit connected by a set of connections. It is used to check the fitness of combatant, spot his position, and detect any hazards in war zone. The obtained signals will be converted into required format using ADC, which is compared with the typical conditional signals. There is a crisis if incongruity between the two occurs.

Chief of Armed Force's Segment This segment has data aggregation facility to collect data from combatant's unit and remove any redundant data. The aggregated data will be sent to the army control unit through cloud using K-means clustering algorithm.

Army Control Room's Segment This segment has a computer system with cloud infrastructure facility. The data available on cloud will be displayed on the computer system.

Figure 9.3 depicts the flow chart of location tracking and updating fitness status of combatant to control room for initiating rescue operation.

Figure 9.4 shows the circuit connection implementation of the project work carried out by interfacing Arduino Uno, NRF Module, and Zigbee.

Results

Following are the few sample results of the project. See Figures 9.5–9.8.

Real-World Use Cases of K-Means Clustering

Following are few interesting use cases of K-means clustering algorithm:

1. **Record Classification**
   It groups archives in various classes depending on labels, subjects, and the substance of the report. This is an extremely

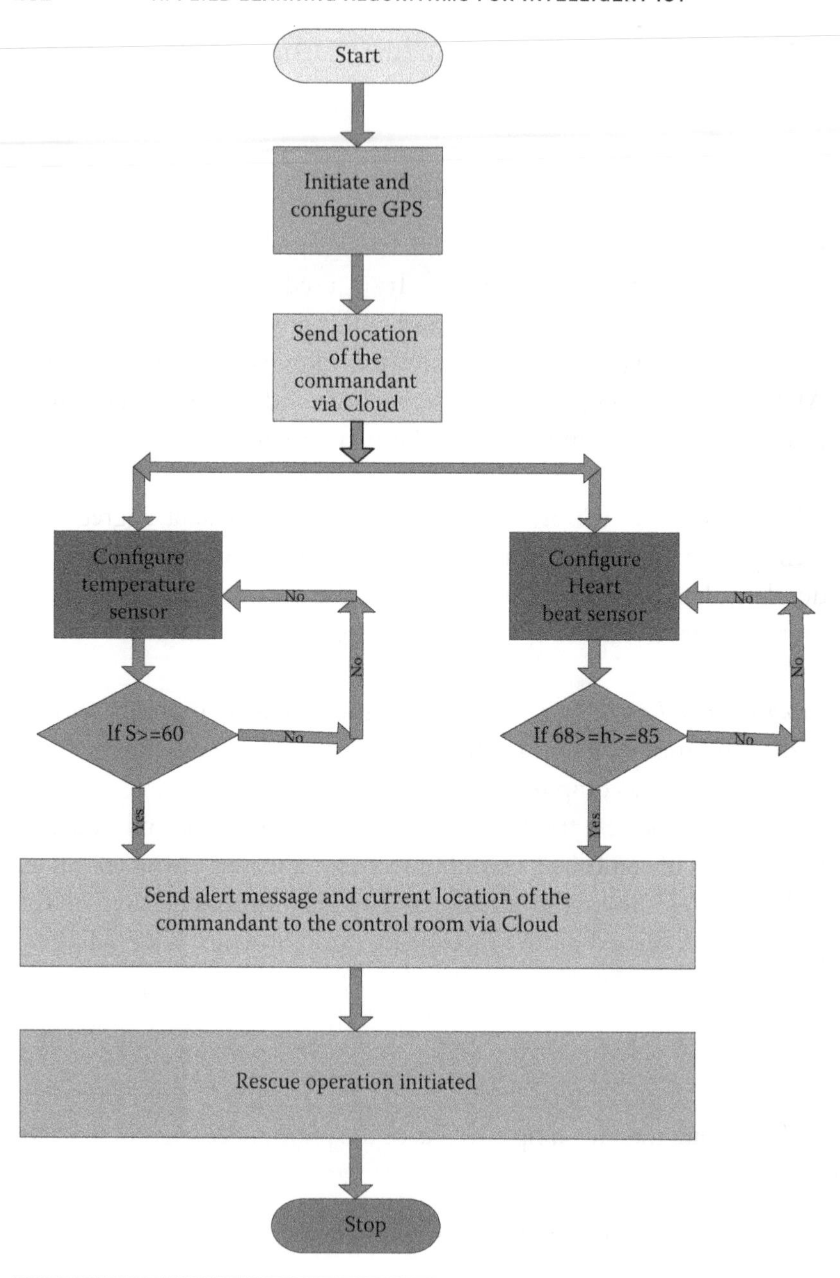

Figure 9.3 Flow of the Proposed Model

Figure 9.4 Circuit of the Project Work

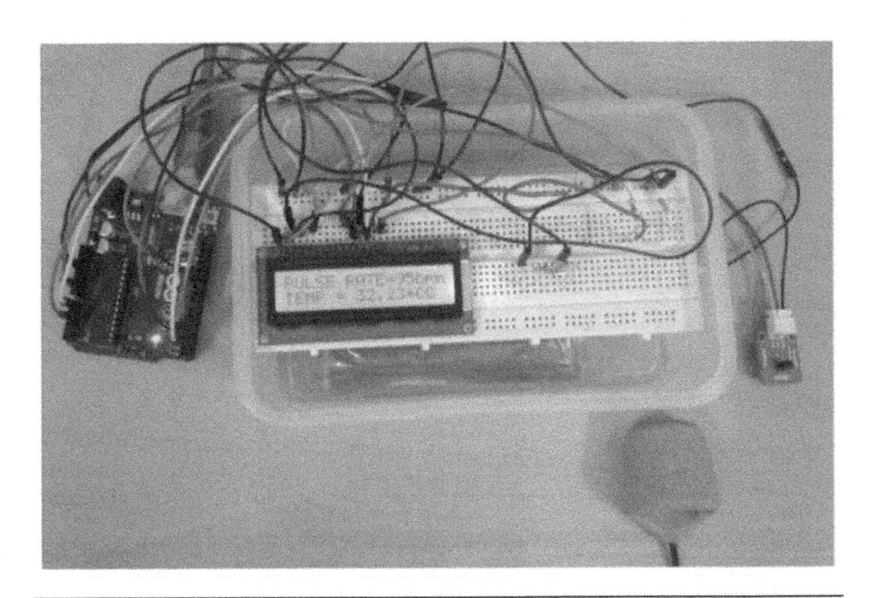

Figure 9.5 Display of Commandant Fitness Parameters Like Pulse Rate and Temperature

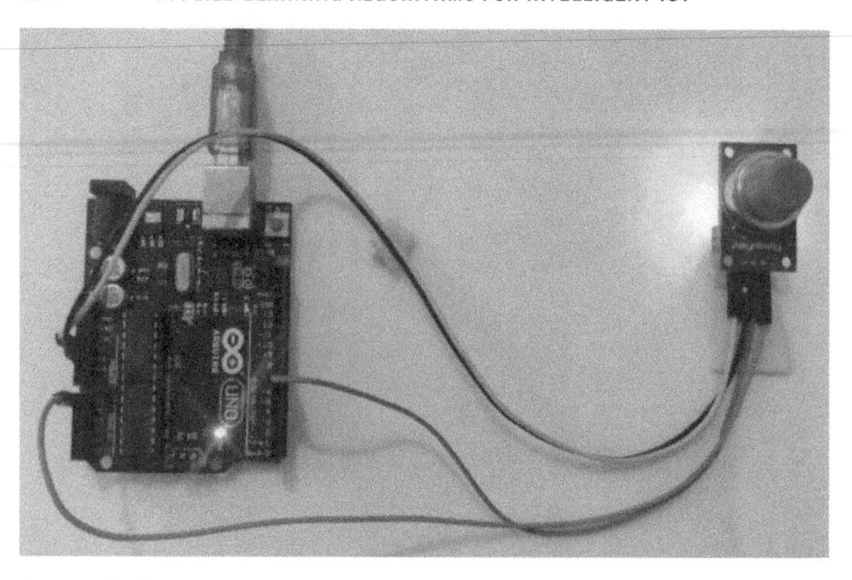

Figure 9.6 Hardware Output to Track Any Hazardous Conditions in Armed Conflict

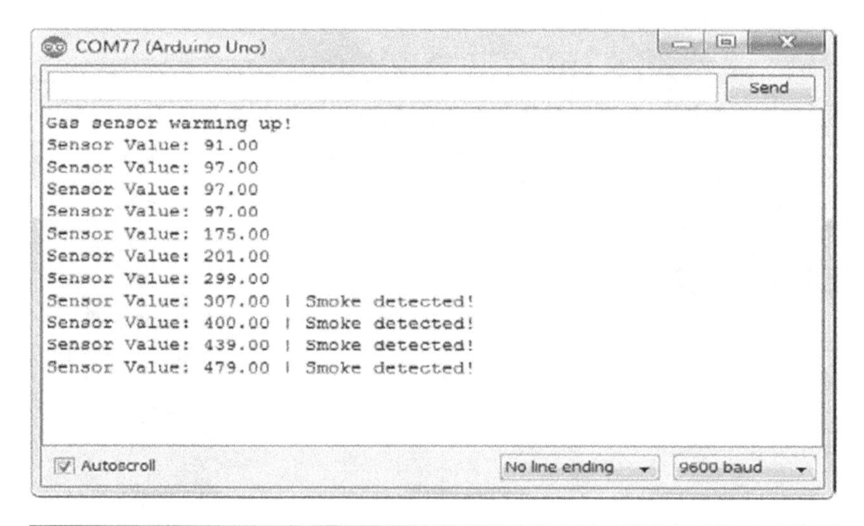

Figure 9.7 Display of Gas Sensor Output to Track War Zone Hazards

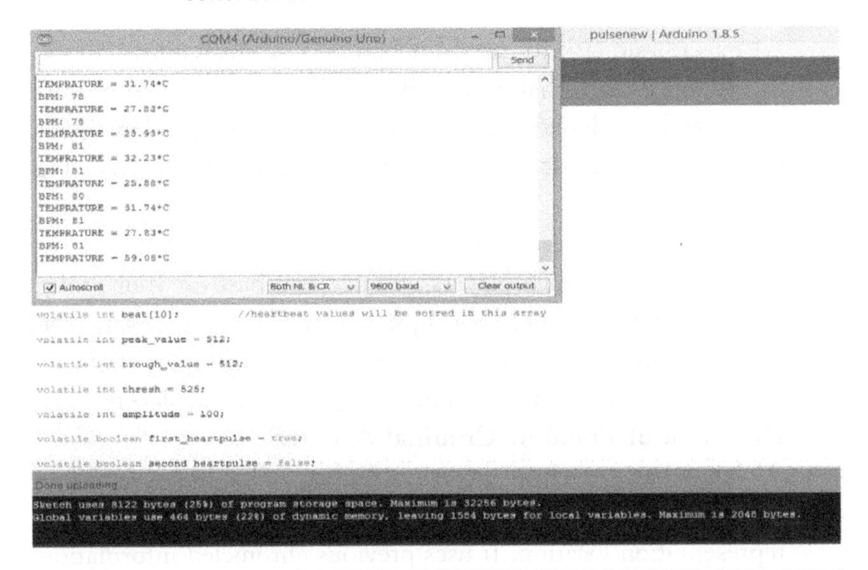

Figure 9.8 Display of Heart Beat Tracker

standard order issue and K-implies is an exceptionally appropriate calculation for this reason. The underlying preparing of the reports is expected to speak to each record as a vector and utilization term recurrence.

2. **Conveyance Store Optimization**
   It improves the procedure of high-quality conveyance utilizing automobile drone by a blend of K-intends to locate the ideal figure of dispatch areas and a hereditary calculation to unravel the automobile course as a voyaging sales rep issue.
3. **Recognizing Offense Localities**
   With information identified with wrongdoings accessible in explicit territories in a city, the class of wrongdoing, the territory of wrong doing, and the relationship between the two can give quality understanding into wrongdoing-inclined zones inside a town or an area.
4. **Client Segmentation**
   Grouping assists advertisers in improving their client database, taking a shot at target regions, and fragmenting clients depending on previous purchase, interests, or movement checking.
5. **Dream Group Statistics Study**
   Breaking down participant details has consistently been a basic component of the donning scene, plus with expanding rivalry,

AI has a basic task to carry out here. As an intriguing activity, on the off chance that you might want to make a dream draft group and similar to recognize comparable troupe dependent on team member details, K-means is capable of be a valuable alternative.

What is the issue: Who would it be advisable for you to have in your group? Which players will perform best for your group and permit you to beat the opposition? The test toward the beginning of the period is that there is less information accessible to assist you with distinguishing triumphant players.

6. **Detection of Fraud or Criminal Activity**
   AI has a basic work in extortion finding. This has several applications in car, medicinal services, and protection misrepresentation location. It uses previous chronicled information on deceitful scenarios, based on its closeness to groups showing fake examples. The ability to recognize cheats is predominant and necessary.

   What is the issue: You need to investigate false driving development. The test is how might you perceive what is valid and which is counterfeit?

   How batching functions: By probing the global positioning system details, the computation can gather proportional practice. Considering the attributes of the gathering, we can exemplify them into authentic and fake.
7. **Rideshare Data Analysis**
   The openly accessible Ola ride dataset gives data on traffic, time of travel, pickup points, etc. Dissecting this information is precious with respect to Ola, as well as in providing knowledge required for smooth travel.
8. **Digital Profiling Criminals**
   Digital profiling is the way toward gathering information from people and gatherings to recognize critical relationship. The possibility of digital profile is acquired from unlawful profile, which gives data on the examination distribution to arrange the sorts of crooks available at the wrongdoing sight.
9. **Call Record Detail Analysis**
   It uses the data caught by telecom organizations during the call, SMS, and web movement of a client. These data give more

noteworthy bits of knowledge about the client's needs when utilized with client socioeconomics. We can bunch client exercises for 24 hours by utilizing the unaided K-implies grouping calculation. It is utilized to comprehend fragments of clients regarding their utilization by hours.

10. **Programmed Grouping of Information Technology Alerts**
    Huge endeavor information technology framework research units create large amount of organized messages. Since these relate to outfitted issue, they should be physically checked for priority during procedure execution. Combination of information can give understanding into classifications of alarms and intervening to fix, and can help in disappointment forecasts.
11. **Identifying Fake News**
    Fake news is not a new phenomenon, but it is one that is becoming prolific.

    What the issue is: Counterfeit reports are being prepared and these reach at a rapid pace for novelty advancements, for instance, online medium. The subject selected was thought as of tardy during the 2016 U.S. official mission. The phrase false report was suggested to an extraordinary amount of epoch.

    The manner wherein the computation works is by captivating the stuff of the not genuine report, the body, looking at the vocabulary utilized, and subsequently gathering them. These gatherings are what assist the calculation with figuring out which pieces are authentic and which are phony information. Few terminologies are exposed usually in sensationalized, misleading content articles.
12. **Spam Channel**
    You know the junk envelope in your email inbox? It is where messages that have been distinguished as spam by the calculation.

    Many AI courses, for example, Andrew Ng's renowned Coursera course, utilize the spam channel to act as an illustration of unaided learning and grouping.

    What the issue is: Spam messages are, best case scenario, an exasperating piece of cutting edge showcasing methods, and even from a pessimistic standpoint, an illustration of individual's phishing for his own information. To try not to get these messages in your fundamental inbox, email organizations use calculations. The reason for these calculations is to signal an email as spam effectively or not.

How grouping functions: K-means bunching strategies have demonstrated to be a compelling method of distinguishing spam. The mechanism is by taking a gander at the various areas of the email. The data are then combined.

Then we have the option to arrange the gatherings to differentiate from spams. Recalling grouping for the gathering cycle improves the accuracy of the channel to 97%. This is a great information for people who are not passing up #1 announcement and offers.

13. **Showcasing and Business Sales**

Marketing and attracting customers in showcasing its product are essential for a large-scale company.

What the issue is: For sales to get the unsurpassed revenue for your advancing speculation, it is vital that you intend folks in the right approach. If you fail to understand the situation, you hazard not doing any deal, or all the further awful, destroying your clients' faith.

Clustering computations can gather together persons with relative attributes and probability to purchase.

14. **Classifying Network Traffic**

Imagine you want to understand the different types of traffic coming to your website. You are particularly interested in understanding which traffic is spam or coming from bots.

What the problem is: As an ever-increasing number of administrations start to utilize APIs on your application, or as your site develops, it is significant you know where the traffic is coming from. For instance, you need to have the option to obstruct tactless traffic and twofold down on zones driving development. Be that as it may, it is difficult to tell which will be with regards to ordering the traffic.

How bunching functions: K-implies grouping is utilized to gather qualities of the traffic sources. At the point when the bunches are made, you would then be able to arrange the traffic types. The cycle is quicker and more exact than the past auto class technique. By having exact data on traffic sources, you can develop your site and plan limit adequately.

15. **Record Examination**

    There are various reasons why you would need to run an examination on a report. In this situation, you need to have the option to sort out the reports rapidly and productively.

    What the issue is: Imagine you are restricted as expected and need to coordinate data held in archives rapidly. To have the option to finish this task, you need to comprehend the topic of the content, contrast it and different archives, and order it.

    How gathering functions: Hierarchical packing has been used to take care of this topic. The estimate can take a gander at the content and assemble it into a variety of matters. Implementing this procedure will help in gathering, as well as coordinating, comparable reports rapidly utilizing the qualities perceived.

K-Means Clustering Algorithm Used as a Solution for Following Business Problems

Purchaser Division

When you have finished group examination, you can recognize portions of the shopper. To do this, you can utilize segment, psychographic and conduct information just as execution information to bunch your purchasers for a specific item class.

Next, you can profile your groups to comprehend your buyers better and depict them as per the factors utilized for bunch examination. You can use this data to tailor your showcasing messages, item collections, and general shopping experience to address your client's issues and increment your ROI.

Conveyance Enhancement

Retailers and providers have hoped to enhance their conveyance procedure utilizing k-implies grouping. The conveyance courses and examples of trucks and automatons have been checked to locate the ideal dispatch areas, courses, and goals for the organization.

Archive Arranging and Gathering

You can aggregate electronic documents as per classification, labels, substance, or recurrence of utilization by K-implies grouping. The calculation sees each archive as a vector and the recurrence of specific terms to arrange and aggregate the records.

Client Maintenance

You can utilize K-implies bunching to break down and bunch client stir to distinguish and profile your buyers depending on maintenance. You can utilize factors, for example, recurrence of buys, how as of late the purchaser visited the store, normal spends per outing. and container arrangement to examine and anticipate degrees of consistency of specific client portions.

Rebate Examination

You can utilize K-implies bunching to scrutinize various gatherings of customers as indicated by their rebate buy practices. Clients might be bound to buy packaged items, items with an ordinary low-value technique, or items at a bargain before expiry. You can utilize the calculation to distinguish buying designs among your customers and use these data to settle on choices with respect to estimating and limiting time methodologies.

Comparisons of K-Means Clustering with Other Unsupervised Algorithms

In this section, a comparison of K-means clustering algorithm with other unsupervised algorithms, namely, hierarchical clustering algorithm (HC), self-organization map algorithm (SOM), and expectation maximization (EM) clustering algorithm is discussed.

Table 9.1 shows comparison of all the above-mentioned four algorithms with respect to performance and accuracy parameters by keeping number of clusters (K) equal to 32.

Table 9.2 shows comparison of all the above-mentioned four algorithms with respect to data types by keeping the number of clusters (K) equal to 32.

Table 9.1 Comparison of SOM, K-Means, EM, and HC Algorithms with Respect to Parameters Performance and Accuracy Keeping K = 32

PARAMETERS	SOM	K-MEANS	EM	HC
Performance	78	84	84	87
Accuracy	830	910	898	850

Table 9.2 Comparison of SOM, K-Means, EM, and HC Algorithms with Respect to Data Types Keeping K = 32

DATA TYPES	SOM	K-MEANS	EM	HC
Random	830	910	898	850
Ideal	798	810	808	829

Table 9.3 Comparison of SOM, K-Means, EM, and HC Algorithms with Respect to Data Size Keeping K = 32

DATA SIZE	SOM	K-MEANS	EM	HC
36,000	830	910	898	850
4000	89	95	93	91

Table 9.3 shows comparison of all the above-mentioned four algorithms with respect to data size by keeping the number of clusters (K) equal to 32.

Pros and Cons

Pros:

1. If K is small and variables are many, then the speed of computation will be more than the hierarchical clustering.
2. If clusters are globular, it will produce clusters that are tighter than the hierarchical clustering.

Cons:

1. Prediction of K is difficult.
2. Diverse preliminary partitions will outcome in diverse concluding clusters.

3. It will not operate well with clusters of dissimilar dimension and dissimilar concentration.

Conclusions and Future Work

Bunch investigation issue has consistently captivated researchers as it bargains with gathering of articles having normal properties. K-means is a basic and adaptable algorithm to trial when you are beginning with grouping. K-means grouping is one of the most popular clustering algorithms and usually the main thing practitioners apply when settling batching tasks to get an idea of the structure of the dataset. The goal of K-means is to aggregate data centers into distinct non-overlapping subgroups.

By and large, K-means calculation is done with Euclidean separation to acquire the groups. K-means calculation is simple to actualize, however, not productive for covering informational indexes. Test perceptions have a place with measurable populaces. One method of tackling this issue is to present factual separation quantifies rather than Euclidean separation.

References

[1] Tajunisha and Saravanan. "Performance Analysis of k-Means with Different Initialization Methods for High Dimensional Data. *International Journal of Artificial Intelligence & Applications (IJAIA)*, vol. 1, no. 4, October 2010.

[2] N. Bouhmala, A. Viken, and J. B. Lonnum. "Enhanced Genetic Algorithm with K-Means for the Clustering Problem." *International Journal of Modeling and Optimization*, 2015, pp. 150–154.

[3] Rouhollah Maghsoudi, Arash Ghorbannia Delavar, Somayye Hoseyny, Rahamatollah Asgari, and Yaghub Heidari. "Representing the New Model for Improving K-means Clustering Algorithm Based on Genetic Algorithm." *The Journal of Mathematics and Computer Science*, 2011, pp. 329–336.

[4] A.P. Jyothi and Usha Sakthivel. "Trends and Technologies Used for Mitigating Energy Efficiency Issues in Wireless Sensor Network." *International Journal of Computer Applications*. vol. 111. 2015, pp. 32–40, 10.5120/19521-1150.

[5]

[6] Sandeep Rana, Sanjay Jasola, and Rajesh Kumar. A Hybrid Sequential Approach for Data Clustering Using K-Means and Particle Swarm Optimization Algorithm. *International Journal of Engineering, Science and Technology*, vol. 2, no. 6, 2010, pp. 167–176.

[7] A. P. Jyothi, and Usha Sakthivel. "CFCLP – A Novel Clustering Framework Based on Combinatorial Approach and Linear Programming in Wireless Sensor Network." In *2017 2nd International Conference on Computing and Communications Technologies (ICCCT)*, pp. 49–54. IEEE, 2017.
[8] Bara'a Ali Attea. A Fuzzy Multi-objective Particle Swarm Optimization for Effective Data Clustering. *Springer*, July 2010, pp. 305–312.
[9] A. P. Jyothi, and S. Usha. (2019). MSoC: Multi-scale Optimized Clustering for Energy Preservation in Wireless Sensor Network. *Wireless Personal Communications*. 10.1007/s11277-019-06146-y.
[10] Chetna Sethi and Garima Mishra. A Linear PCA Based Hybrid K-Means PSO Algorithm for Clustering Large Dataset. *International Journal of Scientific & Engineering Research*, vol. 4, no. 6, June 2013, pp.1559–1566. *International Journal of Computer Applications (0975 – 8887)*, vol. 156, no. 8, December 2016.
[11] Rui Xu and Donald Wunsch. Survey of Clustering Algorithms. *IEEE Transactions on Neural Networks*, vol. 16, no. 3, May 2005.
[12] A. P. Jyothi, and Usha Sakthivel. EPCA: Energy Preservation Using Clustering Approximation in Sensor Network. *ERCICA-2016 Springer Book*, pp. 547–557. 10.1007/978-981-10-4741-1_47.
[13] C. Xiaohui, E. P. Thomas, and P. Paul. Document Clustering Using Particle Swarm Optimization. *Swarm Intelligence Symposium*, pp. 185–191, Pasadena, CA, USA, 2005.
[14] A. P. Jyothi, and S. Usha Interstellar Based Topology Control Scheme for Optimal Clustering Performance in WSN. *Int J Commun Syst.*, vol. 33, 2020, p. e4350. https://doi.org/10.1002/dac.4350.
[15] Osama Abu Abbas, Comparisons between Data Clustering Algorithms. *The International Arab Journal of Information Technology*, vol. 5, no. 3, July 2008.

10

Systematic Approach to Deal with Internal Fragmentation and Enhancing Memory Space during COVID-19

MOHAN APARNA, R. MAHESWARI, AND J. V. THOMAS ABRAHAM

SCOPE, Vellore Institute of Technology, Chennai, Tamil Nadu, India

Contents

DOI: 10.1201/9781003119838-10

Introduction

Internal Fragmentation

The world is jeopardized by technology. With the escalating technology, there is a demand of storage space in the operating system. For the users and operating system developers, the system will run efficiently only if there is an efficient way to manage the memory. There is a frequent need to move and remove the processes from and to the memory. While doing so, there exists the issue of fragmentation. After some point of time, it becomes difficult for the system to accommodate processes into the memory blocks. One of the significant causes for internal fragmentation is the allocation of large memory space for a process that is smaller in size. This leads to empty memory spaces during contiguous memory allocation. Internal fragmentation can be best viewed as the difference between the desired memory size and the size of the memory that is allocated.

Best-Fit Approach and Memory Bank

The main reason for internal fragmentation is the memory block in comparison with the size of the process to be accommodated is bigger. In this scenario, a portion of the memory is left unused and other processes cannot occupy this memory space as well. A memory bank can be created in the memory management system, which would resemble the main memory, but it is a virtual memory that is also divided into blocks. When a process occupies a memory block using best-fit approach, the residual memory can be moved to this memory bank. Accumulation of such residual spaces will be enough for any other process to occupy the space in the memory bank. If the space is found to be adequate to accommodate one process, then that process can be stored in this memory space. Only if the size of the process is the same as the size of the memory block, a process is stored. The remaining space in the memory bank is given to the next block to accommodate another process in the queue. Hence, this proves to be an efficient and optimized way to prevent the wastage of memory space in internal fragmentation and also solves the lack of storage issue. Hence, this research focuses on preventing the wastage of memory space, which occurs in internal fragmentation.

Linear and Lasso Regression

Linear regression is unpretentious and most widely used analytical method for predictive modeling. It provides an equation, where the features are categorized as dependent variables, on which our target variable depends, and independent variables. Lasso regression could also be a kind of linear regression that uses shrinkage. The Lasso procedure encourages sparse models. Lasso means least absolute shrinkage and selection operator. Lasso regression aims at minimizing the prediction errors by implementing certain constraints on the model.

Literature Survey

Dynamic Memory Allocation and Internal Fragmentation

In the field of network and communications, the need for dynamic memory allocation is escalating at a greater pace. According to Stylianos Mamagkakis et al. the real-time operating systems use *state-of-the-art dynamic allocators* [1]. The five types of memory allocators used so far have improved the dynamic memory allocation process but failed to provide a permanent solution for internal fragmentation. The existing memory allocators don't fully serve the purpose of allocating different sizes of memory blocks for the wired and wireless applications in network. This process is prone to internal fragmentation and also causes heavy traffic in run-time memory requests, which is due to its reliance on dynamic memory allocation, as stated by the author Voss M. [2]. Hence, the efficiency, performance, and de-fragmentation techniques of these existing allocators in the world of operating system are sparse [3].

Internal Fragmentation Standards and Compatibility of Systems

Christian Del Rosso states that the compatibility of a system when internal fragmentation occurs is controlled by two significant sources, which are configurational and competitive. When a study was done within the *DVB-H (Digital Video Broadcasting – Handheld)* and *T-DMB (Terrestrial-Digital Multimedia Broadcasting)* mobile digital multimedia broadcasting standards, it was observed that the systems were not compatible with the fragmentation issue that occurred. Techniques like *adoption of a single standard and internal standards fragmentation* were proposed [4]. This was found to be prevalent in the world of mobile-operating systems as well. The issue of compatibility further reduced the performance of the system. Developers found a delay in the loading of addresses and writing to the memory [5]. The compatibility issue was solved using optimized systems and new techniques of memory management in the later stages of development.

Processor Allocation and Internal Fragmentation

Several strategies were introduced to limit the rate of fragmentation in the systems. One such strategy is *Adaptive Non-Contiguous Allocation algorithm.* This works on the basic principle of partition binding and allocating processors. The over-spitting of requests in parallel conditions is also avoided using this algorithm. According to author Bani Ahmad, an optimal value for partition binding is chosen, which will make this algorithm more flexible and tuneable. Using this approach, parallel jobs can be allocated by the processors quickly based on the partitioning bound values [6]. This strategy was found to be compatible with the high loads in communication in comparison to other existing techniques of non-contiguous memory allocation [7]. Contiguous memory allocations fail to reduce the rate of internal fragmentation when larger implementations are considered.

ML and Cloud Storage

The existing machine learning (ML) algorithms used for securing data would categorize a node as an unusual node depending on temporary behavioral data. These systems do not distinguish if an unusual node is a malevolent node or a broken node. Hence for solving this insecurity, the work proposes a *long short-term memory (LSTM)* model [8], which initially learns the behavior of a user and automatically trains itself and stores the behavioral data. The model can easily categorize the user behavior as standard or non-standard. It can also find if a non-standard node is a broken node or a new user node or a tampered node using the calculated trust factor. The proposed technique detects the security attack and safeguards the cloud.

Best-Fit Approach and Internal Fragmentation

According to Rachael Chikoore et al., the genetic algorithms proved as an efficient way to reduce the internal fragmentation caused in the memory management of a system [9]. To produce an optimal configuration, the workload was used as an input. Rational approach was adopted to find the solutions through genetic algorithms [10]. One of the optimized approaches was *best-fit* approach. In this approach, the process size was observed to be greater than the memory size. The internal memory fragmentation was effectively solved using this approach but there was some wastage of memory involved in this process. Though usage of this approach was beneficial, it had its cons when considering evolution, complexity, and adaptability. It also paved way for providing an optimized configuration parameter for a system for free segregated lists. When compared to the historical "brute-force" technique, the best-fit approach was found to be a more heuristic, feasible, and optimal way to deal with internal fragmentation.

Storage and Dependability Study

In today's world, the escalating need for *data storage systems (DSS)* has been turning out as a nightmare as they are attacked by soft errors that are often found in controllers used for storage. The functionalities and stack comprising hardware and software, detection, and correction of errors in DSS stand out when compared to the architectures that follows general-purpose computing. The existing systems or techniques did not address the system-level effects of these soft errors on DSS. This study gives a clear perspective of the effects of the errors in the controllers used for storage and the whole of system dependability is analyzed. In the proposed technique [11], several functions related to storage controllers operating on a full stack of storage system were implemented and a new framework was developed. The impact of soft error in storage systems is accurately calculated using the metric storage system vulnerability factor (SSVF). The results of this analysis helped to draw conclusions that there is data loss when 40% of cache is occupied by user data, and it differs according to the storage controller configuration. It is concluded through the study that data unavailability is as a result of errors which are detectable and cannot be recovered in the cache memory.

Lasso Regression Analysis and Cloud Computing

Cloud computing provides services as and when the need arises for the users. All the resources in cloud are stacked in the data centers, which give rise to optimal resource allocation methods. The existing solutions include greedy algorithms and ML algorithms. As greedy algorithms were NP-hard in nature, it couldn't provide the best solution, but ML provides nearly optimal solution. In the proposed technique [12], ridge regression and lasso regression is implemented on the resource needs of the user dataset. It was concluded that the proposed algorithms gave better results than linear regression because of the correct selection of features and ridge regression has the ability to solve multicollinearity issue. The analysis of the models is done by obtaining root mean squared error (RMSE) and the graphs are plotted. It was observed that the proposed algorithms aided virtual machines to get drain in terms of CPU utilization and storage.

Data Placement and Machine Learning

As cloud computing usage is accelerating, several applications require higher performance, which creates a demand on the storage space available. The data obtained from various sources will contain different frequencies of read/write, retention time, and data of different sizes. A solution to this is hybrid storage system. The underlying issue was the placement of data in hybrid storage based on run-time status and several other properties of data and the system used for storage. In this work [13], *Archivist* – an ML-assisted data placement mechanism for hybrid storage systems – is introduced to minimize the file access latency. To know the access patterns of the input data, ML approach is followed. Furthermore, a data placement algorithm is implemented with an aim to enhance the data on the hybrid storage mediums. This is done by exactly finding out similar properties of data and storage medium features. The experimental results help in drawing conclusions that the performance of the system is improved up to 49% for accessing files compared to baseline.

Memory Resource Management

To increase the performance of a system, the number of cores present should be increased. Several studies indicate that memory management has been a consistent problem which caused a deprecating performance especially in Android devices. The existing techniques focuses on improving efficiency by more allocation of resources and very few studies focus on dealing with internal fragmentation. In the proposed technique, intra-thread behavior [14–34] was focused. In consequence, a *memory management based on thread behavior (MMBT)* was introduced. But it was found that the proposed technique lacked unified optimization program interface and good architecture. To solve this issue, *thread-oriented memory management layer (TOMML)* was proposed. This uses the microkernel architecture pattern and was able to fulfil the user's need for selecting plug-ins to attain diverse optimization goals. Initially the efficiency of TOMML was observed through theoretical simulation and experimentation. As a result, the percentage of memory allocation efficiency was increased by 12–20%. In comparison to previous experimental studies, this technique also saves power by allowing the user to select the plug-in.

Existing Technique

Overview of the Existing Technique with Illustration

Internal fragmentation has been a persistent issue faced in the world of operating system. *Best-fit* approach has been adopted for decades and accepted as the best solution for internal fragmentation. Consider two processes, P1 and P2, of size 25 MB and 30 MB, respectively, as shown in Figure 10.1. The process goes from the secondary memory to the main memory and occupies a block in the main memory. P1 and P2 are accommodated in 1000 MB of space where Process 1 occupies only 25 MB and process P occupies only 30 MB. In each of the above cases, 475 MB of memory space is being wasted in *fragment 1* occupied by P1 and 470 MB of space is being wasted in *fragment 2* where P2 is accommodated. To combat this wastage of space, theories were proposed and researchers found ways to make use of the block of space in the memory which is large enough to accommodate the process but not the largest available memory space. This gave rise to the method followed till date called the Best-Fit approach. In this approach, the process of allocating the

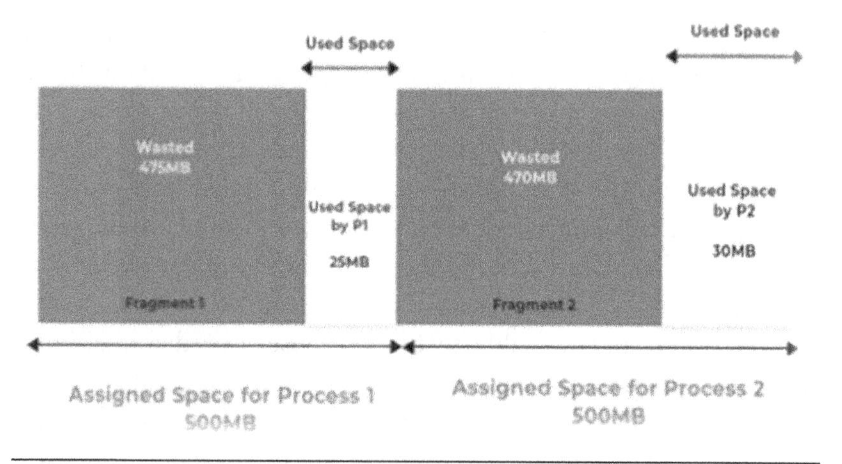

Figure 10.1 Internal Fragmentation

smallest partition but large enough for the process is adopted. However, the problem will not be resolved entirely but can be mitigated to some extent. The illustration of the existing system in Figure 10.1 also resulted in the lack of ability of the system to store the incoming processes in the main memory as a result of substantial or no available memory space. The best-fit approach ensured an immediate but not an optimized solution to the problem of internal fragmentation, faced by the operating system developers for several years. Here, different colors are used for the purpose of illustration to distinguish between the wasted memory space and the block of space allocated to the processes P1 and P2. Hence, this research work provides an optimized approach to avoid this wastage of memory, thereby finding an effective solution for internal fragmentation.

Proposed Technique – Optimized Approach to Deal with Residual Space in Internal Fragmentation

Flow Chart of the Proposed Technique

The flow chart depicted in Figure 10.2 gives a clear perspective of the ptechnique of memory bank approach. Initially, the process is taken as input. When there is a request to move the process from the secondary memory to the main memory, the process first goes to the page table. The Page table maps the logical address and the physical address. The valid and invalid bits are set as “v” and “iv,” respectively, if the page is found and when the page is not found in the memory. In the main memory, the processes are stored according to the best-fit block approach. After implementing the approach, if there is some residual space left in the memory blocks, then that space is accumulated and goes to the memory bank. The memory bank which can be viewed as an instance of main memory can be used to store the incoming processes which cannot be fitted into the main memory. In the memory bank, space is allocated according to the size of the incoming process. This process is repeated and the rest of the process in the queue, waiting for the allotment of a memory block, can be easily stored in the subsequent blocks in the memory bank. The sequential execution of the

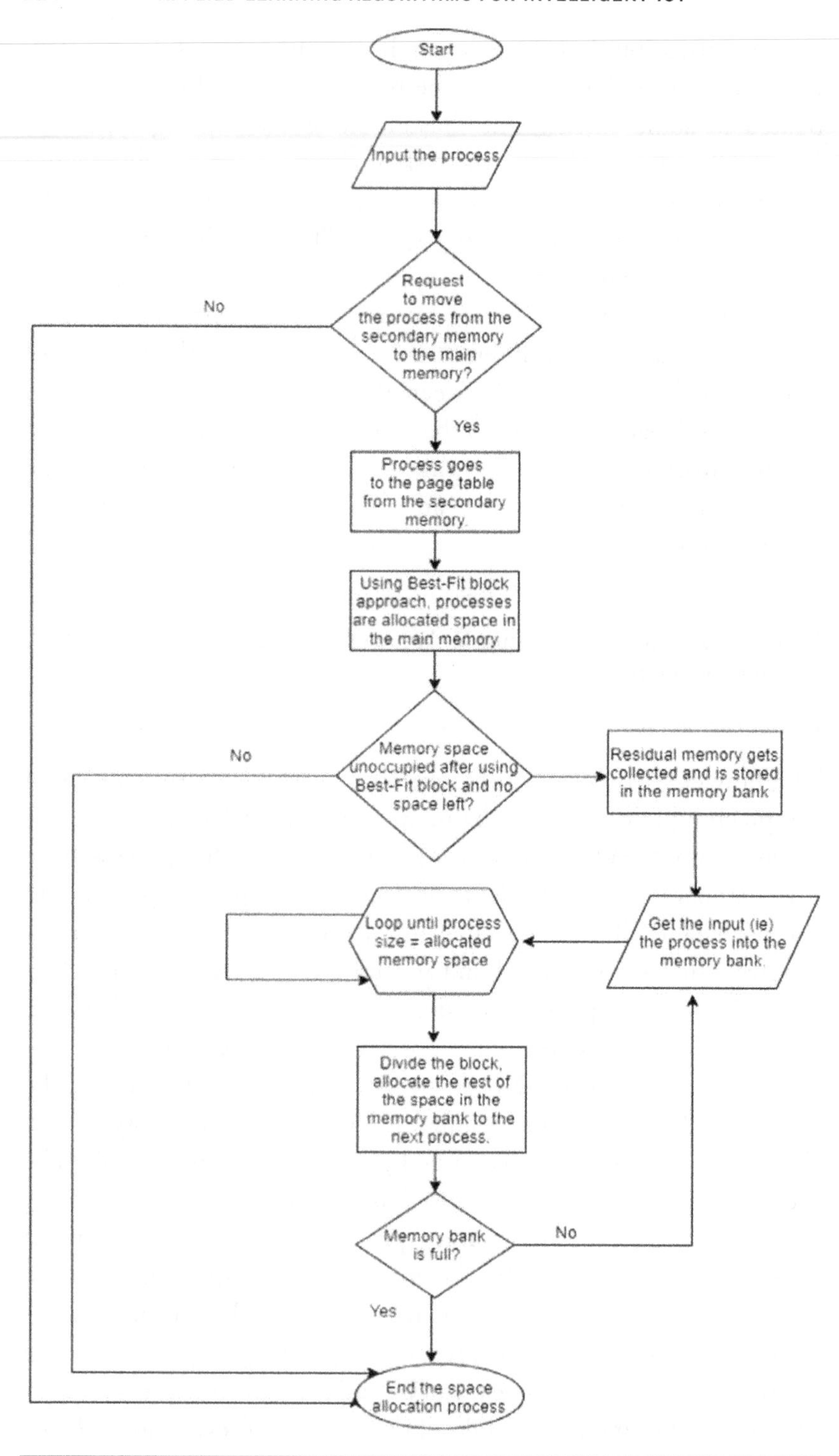

Figure 10.2 Flow Chart Depicting the Proposed Technique

memory bank technique as depicted in Figure 10.2 also enhances the memory management and the general working of a specific operating system. The space allocation in the proposed approach is terminated once the memory bank is full and has no more blocks of space to further accommodate the processes. The residual space computation is an indication of the available space in the memory bank. This helps the developer to determine if the threshold capacity is attained. It also proves to be useful in finding out if there is a possibility to store the further incoming processes in the memory bank. Thus, the technique helps in the prevention of wastage of space after implementing best-fit approach and no process is left behind without allotting a specific memory block.

Schematic of Memory Bank Approach

Memory management plays a significant role for an operating system to function effectively. Initially, when the user intends to move a process from the secondary memory to the main memory, the process moves to the page table. Page table helps in mapping the logical address with the physical address. When a process requests to access the data in the main memory, the operating system stores the mapping of virtual addresses to physical address. The operating system takes the help of Page table to execute this, as it stores the mappings. Each mapping is known as "page table entry"(PTE), which is referred as page number.

The process, if requested, goes to the memory. Here, the processes are prone to undergo fragmentation. In the schematic diagram depicted in Figure 10.3, "internal fragmentation" occurs. Internal fragmentation is viewed as difficult to regain or retrieve compared to other. The process here is larger than the memory space but the fragmentation is solved using the existing "best-fit Approach." During this approach, the process occupies the memory block that is equal to or greater than the size of the process. In Figure 10.3, three processes occupy memory space, which is greater than the size of the process. Hence, there exists residual space in the blocks of main memory. These residual blocks are collected and stored in a replica of the main memory called *memory bank*. Memory

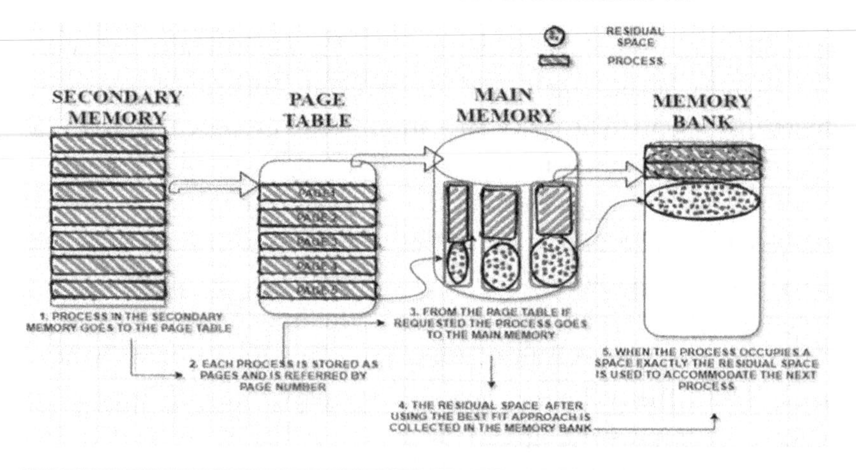

Figure 10.3 The Proposed Technique is Depicted in the Form of Schematic Diagram

bank can be best described as a virtual instance of memory which can hold the processes. These processes can be retrieved as and when needed by the developer. The memory bank depicted here is used to store the residual space which has occurred due to internal fragmentation after implementing the best-fit approach. Once the main memory is full, there occurs a lack of storage issue. Hence, the incoming processes are not able to get stored in the main memory. This memory bank approach can be effectively used to resolve the storage issue as well. The incoming processes which are not able to accommodate in main memory is pushed to this memory bank. Now, the residual space which we have already stored in the memory bank can be used to store these processes. Here once the process size is equal to the memory block size, the storage is stopped and the rest of the space is used for the next incoming process. This approach has now resolved the wastage of memory which prevailed in the existing technique of the best-fit approach in an optimized manner. It has also increased the performance of operating system through effective memory management.

Memory Bank Algorithm

Input: psize_in_GB (Process size), process_name, p_no (Process number), frame_no, vi, memory_block_size

Output: Remaining space in memory bank, graphical representations of Best-fit and Memory bank approaches with process size and process no.

Algorithm 10.1 bestfitfunction(psize_in_GB, memory_block_size, size_process, size_block)

```
bf_allot is set to zeros (1,size_process);
    bf_flag is set to zeros (1,size_block)
    for i=1:size_process
    k is set to -1;
    for i=1:block_size
    if psize_in_GB(i) is less than or equal to memory_block_size(j) &&bf_flag(j)
is 0 then
    if k==-1 then
    k is set to j;
    elseif(memory_block_size(j) is less than memory_block_size(k))
    k is set to j;
    end if
    end if
    end for
    if(k is -1)
    bf_allot(i) is set to 0; // if no block can accommodate the process, set allotted
block number as 0
    else
    bf_allot(i) is set to k; // store the selected index in the best fit array in the index
position i
    bf_flag(k) is set to 1;// set the status of the selected block as busy endfunction
```

Algorithm 10.2 frag_memorybankfunction(psize_in_GB, process_name, p_no, frame_no, vi,memory_block_size, total_size)

```
Print processes in the secondary memory;
    Print page table entries;
    for i =1: n
    if psize_in_GB(i) is equal to memory_block_size(j) then
    Print "process i stored in memory successfully without fragmentation";
    break;
```

```
    elseif psize_in_GB(i) is less than memory_block_size(j) then
    Print"Internal fragmentation occurred memory is allocated using best-fit
approach";
    residual_space is memory_block_size(j)-psize_in_GB(i);
    end elseif
    end if
    end for
    memory_bank_size is residual_space+memory_block_size(i)
    if total_size is greater than 15 then, Greater than RAM size
    Print " RAM size is full. Storage Unavailable! Cannot store process";
    end
    if psize_in_GB(i) is less than or equal to memory_bank_size then
    Print "Process i of size fitted into the memory bank successfully";
    end
    mb_size is memory_bank_size-psize_in_GB(i);
    return residual_space_in_memory_bank;
    endfunction
```

Algorithm 10.3 function display(allot, allotsize, size_process, psize_in_GB)

```
for i=1:size_process, printing process no., process size, block no., block size
    if allot(i) is 0 then
    Print "Not Allocated" with psize_in_GB(i)
    else
    Print(i,psize_in_GB(i),allot(i),allotsize(i))
    end
    end
    endfunction
```

Algorithm 10.4 Main_function

```
size_process is size(psize_in_GB); Determining the number of processes and
blocks
    size_process is size_process(2);
    size_block is size(memory_block_size);
    size_block is size_block(2);
```

To call the function, defined in best fit.sci, for best fit allocation

bf_allot = bestFit(psize_in_GB, memory_block_size, size_process, size_block)

bf_allotsize is zeros(1,size_process), bf_allotsize - size of the selected blocks for best fit

for i=1:size_process, storing the allocated block size for each process

if bf_allot(i) is not 0 then, checking if any block is selected

bf_allotsize(i) is set to memory_block_size(bf_allot(i)), storing the size of the selected block for best fit

else

bf_allotsize(i) is set to 0, store size as 0 if no block is selected

end

end

func = frag (psize_in_GB,process_name,p_no,frame_no,vi,memory_block_size,total_size);

Plot the 2d graph with plot2d() for best-fit and memory bank approach

Linear Regression

Algorithm 10.5

Determine the columns x, y as dependent and independent variables.

regression_model = LinearRegression()

regression_model.fit(x,y)

y_predicted = regression_model.predict(x)

evaluate the model

rmse = mean_squared_error(y, y_predicted)

r2 = r2_score(y, y_predicted)

print 'Slope:', regression_model.coef

print 'Intercept:', regression_model.intercept

print 'Root mean squared error: ', rmse

print 'R2 score: ', r2

Plot the graph with the regression line

Lasso Regression

Algorithm 10.6

Determine the dependent and target variables in the Data frame

model = Lasso(alpha=1.0)

```
model evaluation method:
cv = RepeatedKFold(n_splits=10, n_repeats=3, random_state=1)
evaluate model:
scores = cross_val_score(model, x,y, scoring='neg_mean_absolute_error',
cv=cv, n_jobs= (-1))
force scores to be positive:
scores = absolute(scores)
print 'Mean Absolute Error (MAE):' mean(scores), std(scores)
```

Implementation Using Open Source Software – Scilab and Jupyter (Python)

Pseudo Code of the Proposed Technique

Using Scilab

i. Declares the process size (in GB), process number, and process name and assigns values initially.
ii. Displays the processes in the secondary memory.
iii. Determines the number of processes and blocks and store them in size_process and size_block respectively.
iv. *Best-Fit func()* is defined and called with input parameters as *psize_in_GB, memory_block_size, size_process, size_block.*
v. Declares a best-fit array (*here bf_allot is used*), and a flag array *bf_flag.*
vi. Allocates blocks according to the best-fit approach using *k* as an index variable to indicate the smallest block which can accommodate a process. Initially, this value is set to zero.
vii. If (p_size_in_GB is less than the memory_block_size) then:
update k with the index position of the smaller block
Else set k as the index position of the block.
If no block can accommodate the process, set allotted block number as 0.
viii. Store the selected index in the best-fit array in the index position i (process number) and set memory_block as busy.
ix. Call the function *'internal_frag()'* with input parameters as psize_in_GB,process_name,p_no,frame_no,vi,memory_block_size,total_size.
x. Display the page table entries with valid and invalid bits.

xi. if (process size = = memory block size), process is stored in the main memory successfully.
else if (block size is greater than the process size)
Print ('*Internal Fragmentation has occurred*')
if (memory block size is lesser than process size),
Print('*Process cannot be stored in the memory block*').
xii. Compute the space in the memory bank. This is added with the available space in the memory block that could not accommodate a process.
xiii. Return the space available in the memory bank.
xiv. If incoming process size is equal to or lesser than the size in memory bank, then the process is stored in the memory bank.
xv. The best-fit approach and memory bank approach are plotted as graphs.

Using Jupyter (Python)

i. Load the dataset in the environment.
ii. Select the columns which are of interest in the data frame (i.e.) the dependent variable, in this case revenue and target variable type of industry.
iii. Initialize the linear regression and lasso regression models.
iv. Determine the y_predicted value for each of the models.
v. Evaluate the predicted model. In this case, root mean square error is used for linear regression and mean absolute error for lasso regression is computed.
vi. Print the error values after evaluation of the models and the y-intercept.
vii. Plot the obtained linear regression in the form of a graph with revenue of the company on the x axis and type of industry on the y axis which is a binary value. Along the y axis, "0" denotes small-scale industries using cloud and "1" denotes developed industries using cloud.

Graphical Representation

Inference 1 In Figure 10.4, best-fit block approach is plotted using the open source software Scilab. It is inferred that Processes 1, 2, 3,

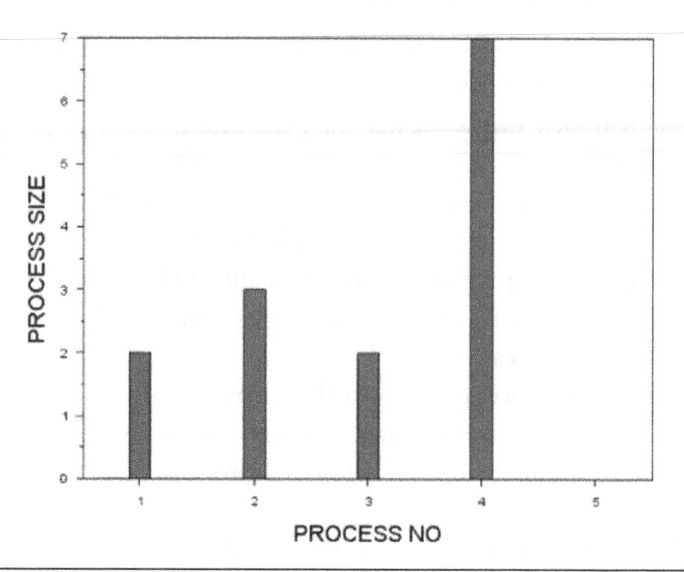

Figure 10.4 Best-Fit Block Approach

and 4 are stored in the main memory according to the size of the processes as shown in the *y* axis. Size of Process 4 is 5 GB but it is stored in the memory block of size 7 GB. The residual space 2 GB is wasted after implementing the best-fit block approach. Process 5 is not stored in the memory as there is no memory block size which is greater than or equal to the size of the process.

Inference 2 From Figure 10.5, it is inferred that the residual space after best-fit block approach is plotted as 2 GB for Process 2. Process 5 which was not plotted in the Figure 10.4 is now plotted as the residual space 2 GB and the available memory block in main memory 1 GB sums up to 3 GB. This memory size will be available in the memory bank and Process 5 is plotted as it is stored in the memory bank. Other processes are plotted with size 0 as they are already stored in the memory using best-fit approach, as seen in Figure 10.5.

Inference 3 From Figure 10.6, it can be inferred that the revenue of an industry also plays as a crucial factor in adopting cloud infrastructure facilities. Here the graph is plotted considering the type of industries as two categories. The developing industries are denoted as "0" and the developed industries as "1" represent along the

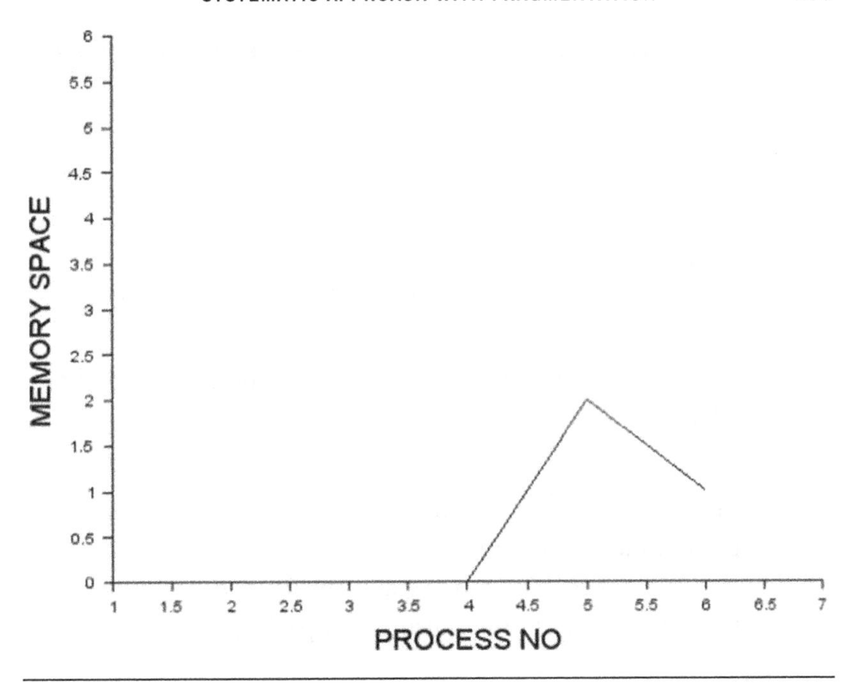

Figure 10.5 Memory Bank Approach Displaying the Residual Space Available

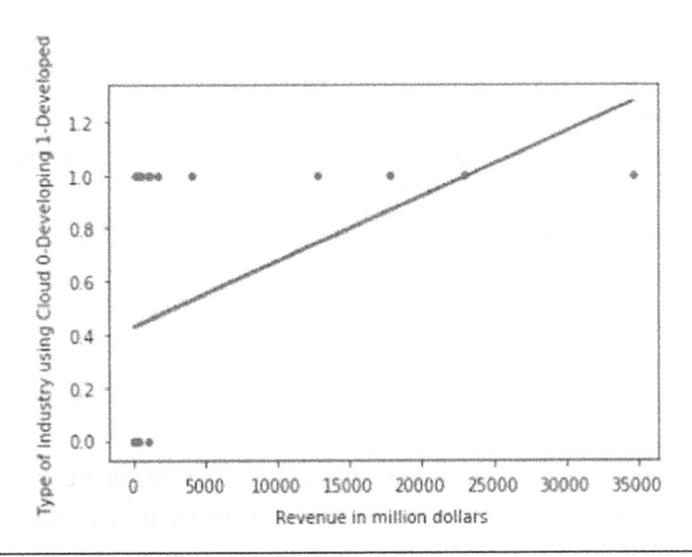

Figure 10.6 Types of Industries Still Lagging in Adopting Cloud Infrastructure

y axis. The revenue of an industry is depicted in million dollars along the *x* axis. From the graph depicted below, it can also be observed that the large-scale companies are adopting cloud infrastructure majorly compared to the developing industries due to the factors like pandemic situation in certain regions, revenue, and size of the industry.

Software Packages Used

1. **Sklearn:**
   - The package sklearn in python helps in implementing many ML algorithms.
   - In this work, modules like train_test_split, DecisionTree-Classifier, and accuracy_score is used.
2. **NumPy:**
   - For faster and efficient arithmetic calculations, the numeric python module NumPy is used.
   - Large number of numpy arrays can be read and manipulations are also performed using this package.
3. **Pandas:**
   - This package is used for the purpose of reading from and writing to different files.
   - Data frames are helpful in performing manipulations.
4. **RStudio – Scatter plot**
   To visualize a scatter plot, this package is used implementing using the function plot (x, y).
5. **Scilab – plot2d**

plot2d is used to plot 2-D curves.

Results and discussions

Figure 10.7 shows the results obtained after implementing the best-fit block approach. The list of processes in the secondary memory, main memory, and page table entries is also displayed. Figure 10.8 gives a clear perspective of the structure of the memory bank. The residual space that is accumulated after best-fit block approach and the space available after the process entering the memory bank are displayed. The obtained results clearly show that

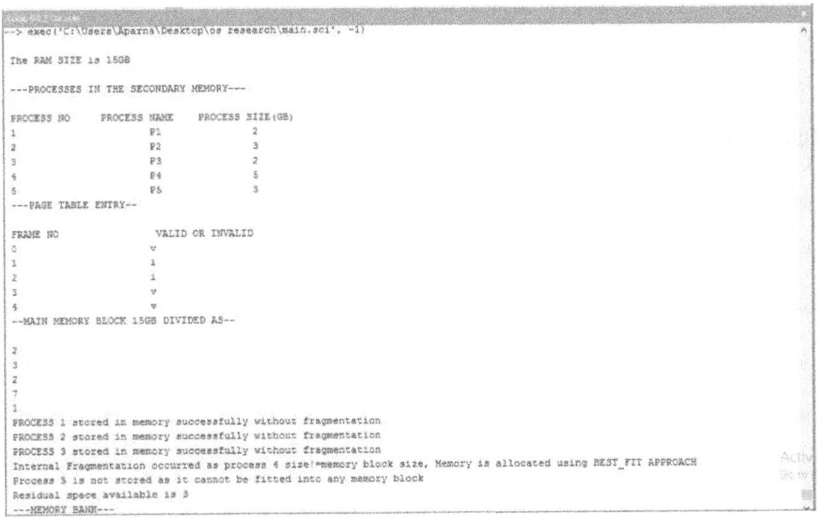

```
--> exec('C:\Users\Aparna\Desktop\os research\main.sci', -1)

The RAM SIZE is 15GB

---PROCESSES IN THE SECONDARY MEMORY---

PROCESS NO      PROCESS NAME    PROCESS SIZE(GB)
1                     P1              2
2                     P2              3
3                     P3              2
4                     P4              5
5                     P5              3
---PAGE TABLE ENTRY--

FRAME NO                VALID OR INVALID
0                       v
1                       i
2                       i
3                       v
4                       v
--MAIN MEMORY BLOCK 15GB DIVIDED AS--

2
3
2
7
1
PROCESS 1 stored in memory successfully without fragmentation
PROCESS 2 stored in memory successfully without fragmentation
PROCESS 3 stored in memory successfully without fragmentation
Internal Fragmentation occurred as process 4 size!=memory block size, Memory is allocated using BEST_FIT APPROACH
Process 5 is not stored as it cannot be fitted into any memory block
Residual space available is 3
---MEMORY BANK---
```

Figure 10.7 Output Displaying Secondary Memory, Main Memory, and Processes after Implementing Best-Fit Block Approach

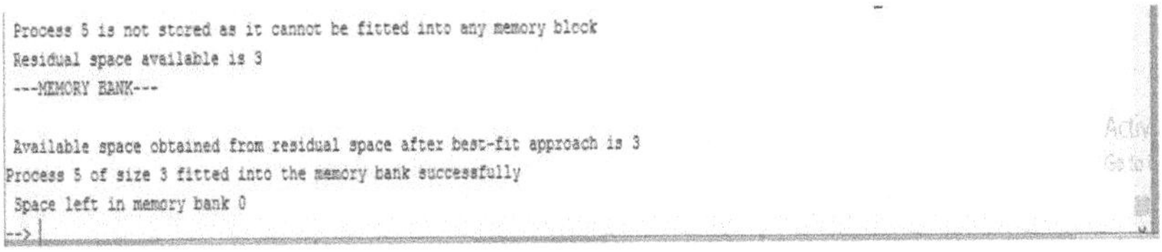

```
Process 5 is not stored as it cannot be fitted into any memory block
Residual space available is 3
---MEMORY BANK---

Available space obtained from residual space after best-fit approach is 3
Process 5 of size 3 fitted into the memory bank successfully
Space left in memory bank 0
-->
```

Figure 10.8 Output Displaying the Memory Bank Approach and the Space in the Memory Bank

the optimized technique is the memory bank approach as no space issue was encountered. The residual space from the best-fit approach was reused effectively to store another process. Around 80% of the number of processes were only stored using the existing approach; hence, the process which is not stored even after using the best-fit approach is stored in memory bank. The space wasted in the previous technique is observed to be 20% which is again reused in the memory bank to store a process which required 20% of the available storage space that is a 3-GB process was stored using the technique of memory bank. Figure 10.9 gives a clear perspective about the root mean square error obtained after prediction of the model using linear regression as 0.195. Figure 10.10

```
Slope: [[2.44442117e-05]]
Intercept: [0.42952059]
Root mean squared error:  0.19535469090977944
R2 score:  0.2106881175362446
```

Figure 10.9 Output Displaying the Results Obtained from the Evaluation of Linear Regression Model

```
-----------------LASSO REGRESSION RESULTS------------
Mean Absolute Error (MAE): 0.478 (0.108)
```

Figure 10.10 Results Obained from Lasso Regression

depicts that overfitting can be greatly avoided using lasso regression as the mean absolute error is computed as 0.478 which is negligible. Hence, the predicted two models are giving efficient results using linear and lasso regression.

Conclusion and Future Work

The problem encountered in the best-fit approach technique is successfully resolved using the proposed technique of using a memory bank. This technique serves as an optimal way to deal with the residual space obtained in internal fragmentation. Moreover, the prevailing issue faced by the developers with the memory management that is the storage issue is also effectively resolved. The memory space in the memory bank is being re-used for the incoming processes, which is not stored in the main memory. The future work of the proposed technique lies in the fact of how far it is used by the developers to enhance memory management in smart phones. It can be further extended to *mobile operating systems (MOS)* to solve the rising sparse in space issues and would serve as the best solution for internal fragmentation. The size of the memory bank can further be extended in the future according to the design of the operating system. The proposed technique also improves the system without any deterioration in the existing data. From the ML predictions, it can be clearly observed that the developing industries, due to various external factors like lack of revenue, location and size of the company during the pandemic situation, are not majorly adopting for cloud infrastructure. Hence, in this COVID-19 situation, the proposed technique of memory bank serves to solve the memory storage issue considering even the small-scale and developing industries.

References

[1] S. Mamagkakis, Christos Baloukas, David Atienza, Francky Catthoor, Dimitrios Soudris, José Manuel Mendias, and Antonios Thanailakis. 2005. "Reducing Memory Fragmentation with Performance-Optimized Dynamic Memory Allocators in Network Applications." In: T. Braun, G. Carle, Y. Koucheryavy, and V. Tsaoussidis (eds) *Wired/Wireless Internet Communications.* WWIC 2005. Lecture Notes in Computer Science, vol. 3510. Berlin, Heidelberg: Springer,.

[2] M. Voss, R. Asenjo, and J. Reinders. "Scalable Memory Allocation." In: *Pro TBB.* Berkeley, CA: Apress, 2019.

[3] Nilesh Vishwasrao and Prabhudev Irabashetti. 2014. Dynamic Memory Allocation: Role in Memory Management." *International Journal of Current Engineering and Technology*, vol. 4, no. 2, April 2014.

[4] Christian Del Rosso. "Reducing Internal Fragmentation in Segregated Free Lists Using Genetic Algorithms." *In Proc. of the 2006 international workshop on Workshop on interdisciplinary software engineering research (WISER '06).* Association for Computing Machinery, New York, NY, USA,. 2006.

[5] D. Atienza, S. Mamagkakis, and F. Catthoor, et al. "Dynamic Memory Management Design Methodology for Reduced Memory Footprint in Multimedia and Wireless Network Applications." In *Proc. of IEEE/ACM DATE*, 2004.

[6] S. Bani-Ahmad. 2011. "Processor Allocation with Reduced Internal and External Fragmentation in 2D Mesh-Based Multicomputers." *Journal of Applied Sciences*, vol. 11, 2011, pp. 943–952.

[7] Lae Wah Htun, Moh Moh Myint Kay, and Aye Aye Cho. 2019. *Analysis of Allocation Algorithms in Memory Management Published in International Journal of Trend in Scientific Research and Development (ijtsrd)*, ISSN: 2456-6470, vol. 3, no. 5, August 2019.

[8] T. Nathezhtha and V. Yaidehi. "Cloud Insider Attack Detection Using Machine Learning." *2018 International Conference on Recent Trends in Advance Computing (ICRTAC)*, Chennai, India, 2018, pp. 60–65, doi: 10.1109/ICRTAC.2018.8679338.

[9] Rachael Chikoore, Okuthe P. Kogeda, and Manoj Lall. 2015. "An Optimized Main Memory Management Dynamic Partitioning

Placement Algorithm." *Pan African Conference on Science, Computing and Telecommunications (PACT)*, July 27–29, Kampala, Uganda, 2015.

[10] D. Koch, C. Beckhoff, and J. Teich. 2009. "Minimizing Internal Fragmentation by Fine-Grained Two-Dimensional Module Placement for Runtime Reconfigurable Systems." *17th IEEE Symposium on Field Programmable Custom Computing Machines*, Napa, CA.

[11] M. Kishani, M. Tahoori, and H. Asadi, "Dependability Analysis of Data Storage Systems in Presence of Soft Errors." In: *IEEE Transactions on Reliability*, vol. 68, no. 1, pp. 201–215, March 2019, doi: 10.1109/TR.2018.2888515.

[12] Harshala Shingne, Sehar Sountharrajan, M. Karthiga, and Rajan Chinnasamy. "Lasso and Ridge Regression for Optimized Resource Allocation in Cloud Computing." *Journal of Advanced Research in Dynamical and Control Systems*, vol. 12. 2020, pp. 1740–1747. 10. 5373/JARDCS/V12I2/S20201215.

[13] J. Ren et al. "Archivist: A Machine Learning Assisted Data Placement Mechanism for Hybrid Storage Systems." In: *2019 IEEE 37th International Conference on Computer Design (ICCD)*, Abu Dhabi, United Arab Emirates, 2019, pp. 676–679, doi: 10.1109/ ICCD46524.2019.00098.

[14] Z. Zhu, F. Wu, J. Cao, X. Li, and G. Jia. "A Thread-Oriented Memory Resource Management Framework for Mobile Edge Computing." *IEEE Access*, vol. 7, pp. 45881–45890, 2019, doi: 10. 1109/ACCESS.2019.2909642.

[15] D. Atienza, S. Mamagkakis, et al. "Modular Construction and Power Modelling of Dyn. Mem. Managers for Embedded Systems." In: *Proc. of LNCS PATMOS*, 2004.

[16] C. Williamson. "A Tutorial on Internet Traffic Measurement." *Proc. of IEEE Internet Computing*, vol. 5, no. 6, p. 70–74,2001

[17] Eran N. Ben-Porath. "Internal Fragmentation of the News." *Journalism Studies*, vol. 8, no. 3, 2007.

[18] M. Norton. "*A Structural Hermeneutics of The O'Reilly Factor*." *Theory and Society*, vol. 40, p. 315, 2011.

[19] Joseph Yiu. 2015. *The Definitive Guide to Arm® Cortex®-M0 and Cortex-M0+ Processors* (2nd Edition).

[20] M. S. Johnstone and P. R. Wilson. "The Memory Fragmentation Problem: Solved?" *Proceedings of the 1st International Symposium on Memory Management*, vol. 34, no. 3, October 1998.

[21] Z. Zhu, F. Wu, J. Cao, X. Li, and G. Jia. 2019. "A Thread-Oriented Memory Resource Management Framework for Mobile Edge Computing." *IEEE Access*, vol. 7.

[22] Durgesh Raghuvanshi. "Memory Management in Operating System." *International Journal of Trend in Scientific Research and Development (ijtsrd)*, ISSN: 2456-6470, vol. 2, no. 5, August 2018.

[23] Amanjot Kaur Randhawa, and Alka Bamotra. "Study of Static and Dynamic Memory Allocation." *International Journal of Innovative Computer Science & Engineering*, vol. 4, no. 3,pp. 127–131, 2017.
[24] Meenu, Vinay Dhull, and Monika. 2015. "Computational Study of Static and Dynamic Memory Allocation." *International Journal of Advanced Research in Computer Science and Software Engineering*, vol. 5, no. 8.
[25] Dharmender Aswal, Krishna Sharda, and Mahipal Butola. "DMA: Dynamic Memory allocation." *IJIRT*, vol. 1, no. 6, 2014.
[26] Dipti Diwase, Shraddha Shah, Tushar Diwase, and Priya Rathod. 2012. "Survey Report on Memory Allocation Strategies for Real Time Operating System in Context with Embedded Devices." *IJERA Internet Computing [Online]*, vol. 2, no. 3, pp. 1151–1156.
[27] Dinesh Kumar and Mandeep Singh. 2019. "Memory Management in Operating System." *International Journal of Emerging Technologies and Innovative Research*, 6, 465–471.
[28] Tie Qiu, Heyuan Wang, Keqiu Li, Huan Sheng, Arun Kumar, Sangaiah, et al. "A Novel Machine Learning Algorithm for Spammer Identification in Industrial Mobile Cloud Computing," *IEEE Transaction on Industrial Informatics*, vol. 15, no. 4, pp. 1, April 2019.
[29] J. Miranda, N. Makitalo, J. Garcia-Alonso, J. Berrocal, T. Mikkonen, C. Canal, et al. "From the Internet of Things to the Internet of People." *IEEE Internet Computing*, vol. 19, no. 2, pp. 40–47, 2015.
[30] T. K. Rodrigues, K. Suto, H. Nishiyama, J. Liu, and N. Kato. "Machine Learning Meets Computation and Communication Control in Evolving Edge and Cloud: Challenges and Future Perspective." *IEEE Communications Surveys & Tutorials*, vol. 22, no. 1, pp. 38–67, First quarter 2020, doi: 10.1109/COMST.2019.2943405.
[31] V. Sze, Y.-H. Chen, T.-J. Yang, and J. S. Emer. "Efficient Processing of Deep Neural Networks: A tutorial and Survey." *Proceedings of the IEEE*, vol. 105, no. 12, pp. 2295–2329, 2017.
[32] S. Gu, Q. Zhuge, J. Yi, J. Hu, and E. H.-M. Sha. "Data Allocation with Minimum Cost under Guaranteed Probability for Multiple Types of Memories." *Journal of Signal Processing Systems*, vol. 84, no. 1, pp. 151–162, 2016.
[33] X. Chen, E. H.-M. Sha, Q. Zhuge, W. Jiang, J. Chen, J. Chen, et al. "A Unified Framework for Designing High Performance in Memory and Hybrid Memory File Systems." *Journal of Systems Architecture*, vol. 68, pp. 51–64, 2016.
[34] X. Zhang, D. Feng, Y. Hua, and J. Chen. "A Cost-Efficient NVM-Based Journaling Scheme for File Systems." In: *2017 IEEE International Conference on Computer Design (ICCD)*, pp. 57–64, 2017.

Appendix 10.A

The source for the dataset used for the ML prediction model is Google Cloud platform and Enlyft online platform as shown below:

Bonobo	Australia	50	10	0
Focus Pointe Global	USA	1000	100	0
Robert W Woodruff Arts Centre Inc	USA	5000	100	1
America's Foundation Inc	USA	200	1000	0
Burroughs Inc	USA	5000	1000	1
20th Century Fox	USA	2300	1091	1
Paypal	USA	21800	17772	1
Team Computers	India	5000	10	0
Indiamart	India	3150	4035	1
Dr Reddy's Laboratories	India	21000	270	1
Truecaller	Sweden	500	200	0
LIC Housing Finance	India	2103	12770	1
Tata Consultancy Services	India	4,48,464	23000	1
DB Corp Ltd	India	11000	320	0
Innoplexus AG	India	210	15	0
Hero Motocorp Ltd	India	8599	34648	1
ShareThis	USA	120	4	0
UbiSoft	France	14000	495	1
Snapchat	USA	3195	1700	1
Zenly	France	69	35	0

11

IoT Automated Spy Drone to Detect and Alert Illegal Drug Plants for Law Enforcement

GOTLURU ARUN KUMAR, PULUGU YAMINI, AND R. MAHESWARI

SCOPE, Vellore Institute of Technology, Chennai, Tamil Nadu, India

Contents

DOI: 10.1201/9781003119838-11

Introduction

In this era, there is no efficient way of detecting the psychoactive drugs without manpower. Here the spy drone has a unique feature that identifies the psychoactive drugs, as it contains an autopilot to identify the location of the drug plants with the help of GPS, which can be explicitly controlled remotely. The collected information will be stored in the database managed by the drone controller. This paper discusses the work on autopilot, image classification techniques to identify the drug plants, and the information security concepts used in developing the remote interface for securing and storing the collected information in the database by the controller. A drone is passed through various areas of the city to see the people roaming around. If the drone controller identifies unwanted gathering or roaming, the location and the information will be sent to the nearest police station. Now when it comes to the proposed work, based on the existing work, the proposed system has been modified accordingly to identify the psychoactive drugs with the help of a drone in which image classification techniques using a python programming language to classify the drug plants in an area are used.

GPS tracker is used to track the location of the identified drug plants by the drone. The shortest path algorithm is used to send the drone to the various locations and get back to the controller. This particular algorithm is used for low consumption of energy and time. Cryptography techniques like vigenere cipher and MD5 hashing algorithms are used for data encryption in the database. Immediately, the police will take the action. An automated spy drone is designed to reveal any persons/groups planting and growing psychoactive drugs unofficially in their private land and identify them. Whenever the spy drone system identifies these psychoactive drugs, it immediately delineates the inspection officer by storing the information in the database automatically.

Objectives

This paper aims to develop an eco-friendly smart spy drone that can be used for large-scale sectors handled by the government to detect the psychoactive drugs and can help in decreasing the growth of illegal drug production. The controller will control the drone with a remote controller by sending the drone to specific locations to identify the psychoactive drug plants. Drone detects the psychoactive drugs grown in private lands and forest areas with the help of image classification techniques. VGG16 network architecture has been implemented for the image classification by the drone. Shortest Path algorithm is used for sending the drone to different locations and to return to the controller. It helps in saving energy and time consumption. As the controller will be monitoring the location details of the drone and the images captured by the drone, he or she will store that information in a database provided in the remote controller. The collected/stored information will be encrypted using cryptography techniques to avoid data leakage. Vigenere cipher and MD5 hashing techniques are used for the encryption. After storing the information in a database, the controller sends the collected information to the head office for further actions.

Literature Survey

In the literature survey, many articles had been referred to and had some research on them. The working principles of VGG16 algorithm in Keras are discussed in Aized Amin Soofi et al.'s work [1]. Cryptography techniques like enhanced vigenere cipher for data encryption were presented by Aized Amin Soofi, Irfan Riaz, Umair Rasheed in Wang et al.'s work [2], hashing algorithm such as MD5 by Shweta et al. [3]. Working and possibilities of drone were discussed by Piotr and Jacec [4] and working and functioning of VGG16 and VGG19 by [5]. Deep convolution networks for large-scale images recognition were discussed by Karen Simonyan and Andrew Zisserman [6].

Existing System

In this pandemic situation, drone technology is used for regular inspection whether people are roaming during a lockdown or not. A drone is passed through various areas of the city to see the people roaming around. If the drone controller identifies unwanted gathering or roaming, the location and the information will be sent to the nearest police station. Immediately, the police will take the action.

Proposed System

Based on the existing work, the proposed system has been modified accordingly to identify the psychoactive drugs with the help of a drone in which image classification techniques using a Python programming language to classify the drug plants in an area are used. GPS tracker is used to track the location of the identified drug plants by the drone. The shortest path algorithm is used to send the drone to the various locations and get back to the controller. This particular algorithm is used for the low consumption of energy and time. Cryptography techniques like vigenere cipher and MD5 hashing algorithms are used for data encryption in the database.

Architectural Diagrams

Figure 11.1 elaborates the complete architecture of remote interface. This interface is visualized and controlled by the controller where the controller can register and login into the interface and this interface provides services like Add, List, and Search.

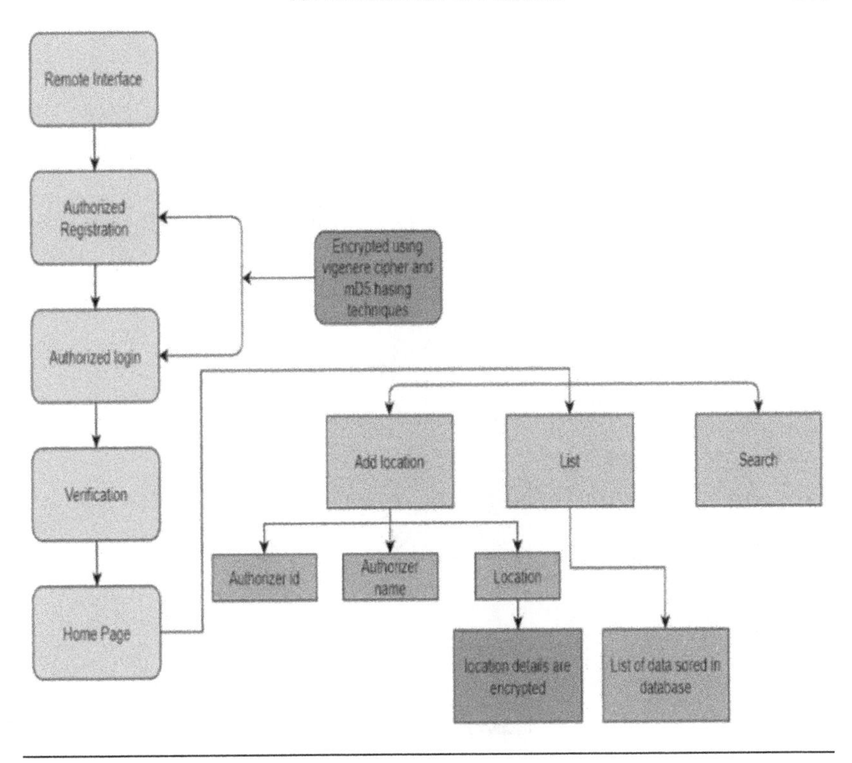

Figure 11.1 Remote Interface Architecture

The authorized person can add the details of the identified plants and location into the database and if admin wants to see the location, then he/she has to authenticate and retrieve data. The login details and location are encrypted using vigenere cipher [2] and hashing technique like MD5 [3] algorithms are used for security. In the overall architecture of the spy drone, there are mainly five modules: The first module is the communication module which is also a database used for the data transmission from drone to the database. The second module is vision sensor. It helps the drone for visualizing the specific location and after the visualization, the identified plants will be classified into normal or psychoactive plants. The next module is mission controller which involves a battery and the remote controller. This remote controller is used to operate the drone by the controller. Drone autopilot is the fourth module that is the drone with an autopilot which has an auto mechanism to get away from distracting objects. The last module is the GPS module that is used to identify the locations and navigate the drone to the particular location. The overall architecture of the spy drone is shown in Figure 11.2.

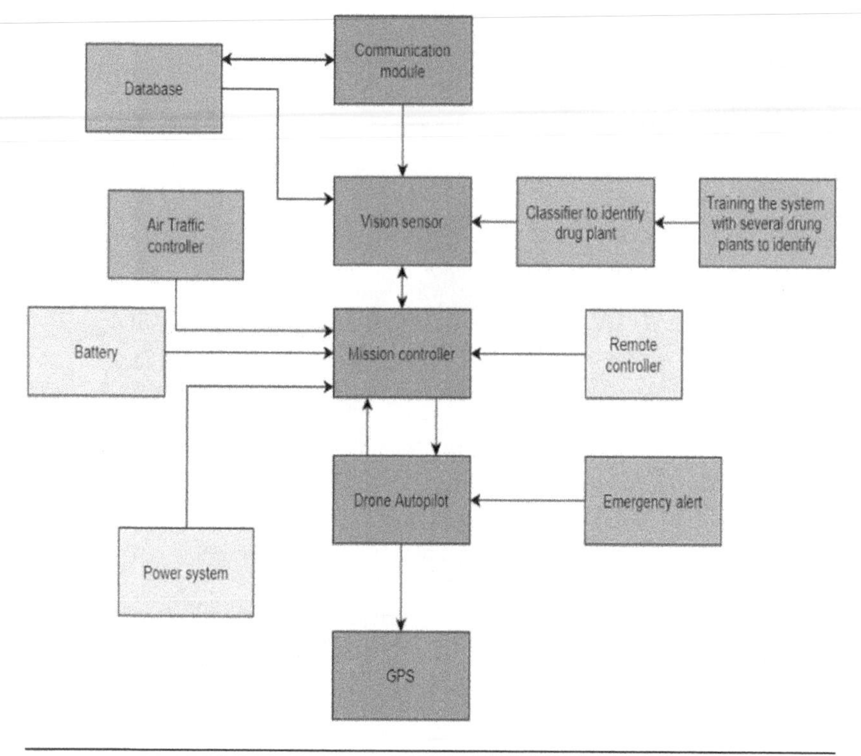

Figure 11.2 Architectural Diagram of Spy Drone

Methodology

Technical Approach

For image classification and identifying plants, ML/Dl algorithms are used [7]. GPS tracker is used to tracking the location of the drone and identified plants. The shortest path algorithm is used to get back to the controller. This particular algorithm is used for low consumption of energy and time. Cryptography techniques like vigenere cipher and MD5 hashing techniques [5] are used for encrypting the collected data to avoid data leakage.

The main steps involved in the machine learning (ML) model are:

- Importing of necessary packages
- Importing of data set path
- Data prepossessing
- Data augmentation

- Splitting of data
- Building of the model
- Training of the model
- Results of the accuracy

Study of each step of the ML model

For the image classification in Spy drone, all the necessary packages are imported from Keras libraries in python. The next step is importing the data set. The dataset will be stored in the respective directory where the notebook file is stored. The path of the dataset directory will be given for importing the dataset for further steps. The collected data don't need to be in the desired format. Therefore, data prepossessing techniques are used. Data prepossessing is a process of preparing the raw data and making it suitable for a machine learning model. For performing data prepossessing using python, we need to import some predefined Python libraries, i.e., Pandas. Data prepossessing techniques include formatting of data, cleaning of data, and sampling of data. Data augmentation is an integral process in deep learning. In deep learning, we need large amounts of data and in some cases, it is not feasible to collect thousands or millions of images, so data augmentation comes to the rescue. There is no need to collect new data; rather it transforms the existing data into the required forms and stores as augmented data. Data augmentation operations include boundary boxing, max pooling, cropping, and extracting the region of interest (ROI).

VGG16 Architecture

It is a convolution-based neural network model integrated with the image net processing [6]. To date in the domain of visual architecture, VGG16 plays a vital role in its excellent performance. The most important and unique feature in VGG16 is that it will not use any hyper-parametric values in it. Rather it will have the convolutional layers and the same type of max-pooling and padding will be processed in this VGG16 modeled architecture. 3 × 3 layered filters will be used in convolutional layers and 2 ×2 layered filters will be used in max-pooling and padding for every time. Activation

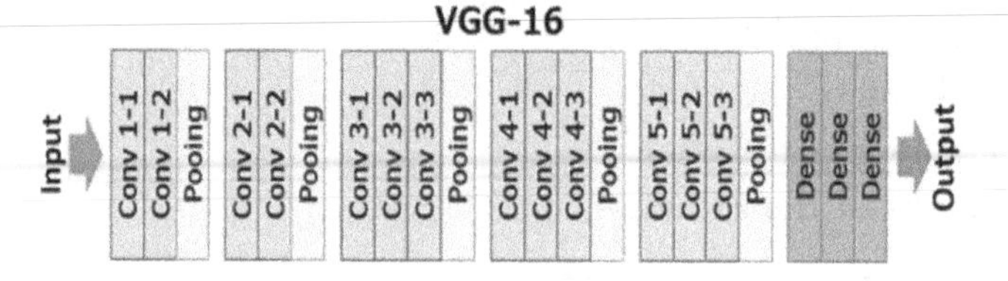

Figure 11.3 VGG16 Architecture

functions like Relu or Softmax will act as fully connected layers in this VGG16 architecture for providing output. In VGG16 [5], 16 represents the weights associated with the model architecture. VGG16 has approximately 138m parameters in it, and this concludes that the VGG16 neural network architecture is a very large network model, as [8] represented in Figure 11.3.

Modules Involved

1. **Image Classification**
   Captured image gets classified with the pre-trained model VGG16. Model has been trained with several dataset plants and it identifies the target plant.
2. **Remote Control Interface**
   Remote controller contains an interface that contains features like Add location, List, and search in database; thus, the person who is going to access/control the drone gets authenticated if he or she tries to get data and see the original data stored in it.
3. **Drone Frame**
   Drone frame has been designed using the tinker-card tool, which is the best tool for virtual simulation of IoT devices.

Description of Working Modules

The Steps Involved in the Image Classification VGG16 Model Import of necessary packages, import of data set, image preprocessing, segmentation, and computer vision model are represented in Figure 11.4.

Figure 11.5 represents the steps of various packages that were downloaded from Keras, how dataset gets imported into the model and visualizes the input images. Next image processing is done by importing the target image and visualizing the image in the RGB

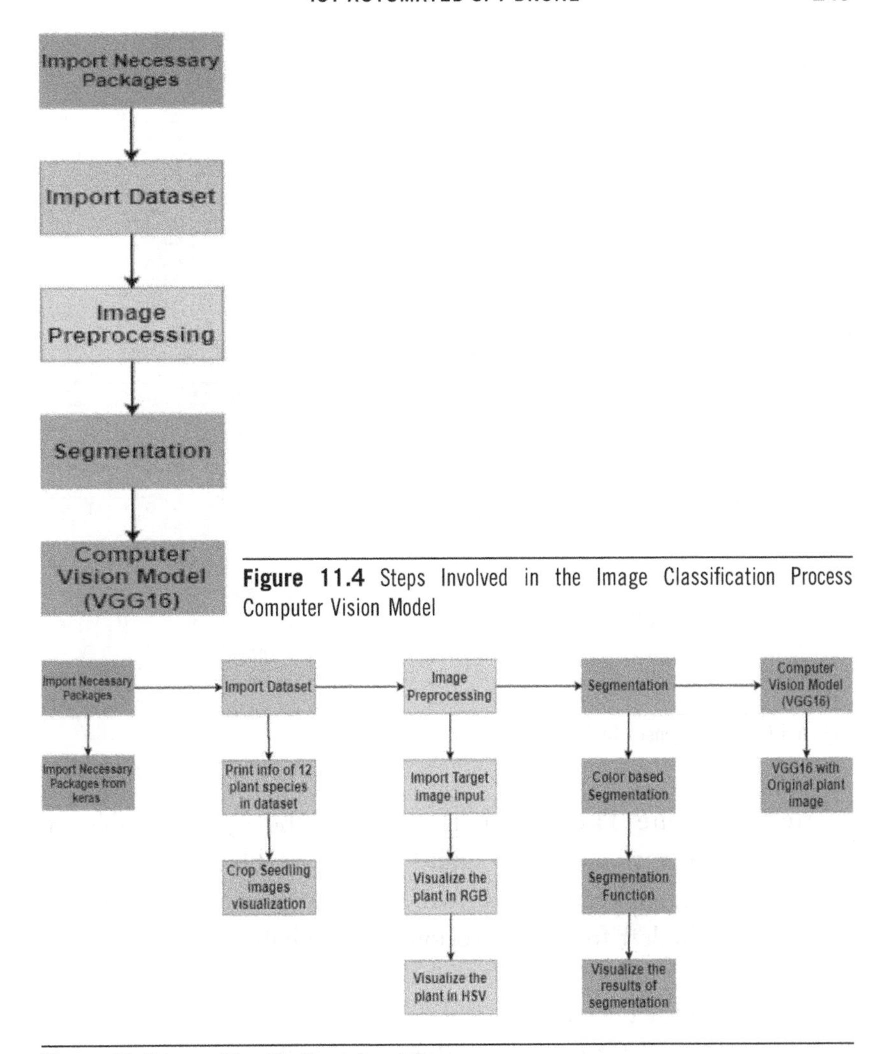

Figure 11.4 Steps Involved in the Image Classification Process Computer Vision Model

Figure 11.5 Image Classification Internal Process

and HSV format; thereby image gets segmented based on color and visualizes the segmentation results in VGG16 model.

Remote Interface Encryption Is Done by Using Vigenere Cipher and MD5 Hashing Techniques

Vigenere Cipher It is a cryptography-based method which is used for

encrypting the text in alphabetical form. The form of poly-alphabetic substitution is used in this method. This cipher is based on the substitution of multiple alphabets. The plain text message will be encrypted for generating the ciphertext using the vigenere table

Plaintext

Key

	A	B	C	D	E	F	G	H	I	J	K	L	M	N	O	P	Q	R	S	T	U	V	W	X	Y	Z
A	A	B	C	D	E	F	G	H	I	J	K	L	M	N	O	P	Q	R	S	T	U	V	W	X	Y	Z
B	B	C	D	E	F	G	H	I	J	K	L	M	N	O	P	Q	R	S	T	U	V	W	X	Y	Z	A
C	C	D	E	F	G	H	I	J	K	L	M	N	O	P	Q	R	S	T	U	V	W	X	Y	Z	A	B
D	D	E	F	G	H	I	J	K	L	M	N	O	P	Q	R	S	T	U	V	W	X	Y	Z	A	B	C
E	E	F	G	H	I	J	K	L	M	N	O	P	Q	R	S	T	U	V	W	X	Y	Z	A	B	C	D
F	F	G	H	I	J	K	L	M	N	O	P	Q	R	S	T	U	V	W	X	Y	Z	A	B	C	D	E
G	G	H	I	J	K	L	M	N	O	P	Q	R	S	T	U	V	W	X	Y	Z	A	B	C	D	E	F
H	H	I	J	K	L	M	N	O	P	Q	R	S	T	U	V	W	X	Y	Z	A	B	C	D	E	F	G
I	I	J	K	L	M	N	O	P	Q	R	S	T	U	V	W	X	Y	Z	A	B	C	D	E	F	G	H
J	J	K	L	M	N	O	P	Q	R	S	T	U	V	W	X	Y	Z	A	B	C	D	E	F	G	H	I
K	K	L	M	N	O	P	Q	R	S	T	U	V	W	X	Y	Z	A	B	C	D	E	F	G	H	I	J
L	L	M	N	O	P	Q	R	S	T	U	V	W	X	Y	Z	A	B	C	D	E	F	G	H	I	J	K
M	M	N	O	P	Q	R	S	T	U	V	W	X	Y	Z	A	B	C	D	E	F	G	H	I	J	K	L
N	N	O	P	Q	R	S	T	U	V	W	X	Y	Z	A	B	C	D	E	F	G	H	I	J	K	L	M
O	O	P	Q	R	S	T	U	V	W	X	Y	Z	A	B	C	D	E	F	G	H	I	J	K	L	M	N
P	P	Q	R	S	T	U	V	W	X	Y	Z	A	B	C	D	E	F	G	H	I	J	K	L	M	N	O
Q	Q	R	S	T	U	V	W	X	Y	Z	A	B	C	D	E	F	G	H	I	J	K	L	M	N	O	P
R	R	S	T	U	V	W	X	Y	Z	A	B	C	D	E	F	G	H	I	J	K	L	M	N	O	P	Q
S	S	T	U	V	W	X	Y	Z	A	B	C	D	E	F	G	H	I	J	K	L	M	N	O	P	Q	R
T	T	U	V	W	X	Y	Z	A	B	C	D	E	F	G	H	I	J	K	L	M	N	O	P	Q	R	S
U	U	V	W	X	Y	Z	A	B	C	D	E	F	G	H	I	J	K	L	M	N	O	P	Q	R	S	T
V	V	W	X	Y	Z	A	B	C	D	E	F	G	H	I	J	K	L	M	N	O	P	Q	R	S	T	U
W	W	X	Y	Z	A	B	C	D	E	F	G	H	I	J	K	L	M	N	O	P	Q	R	S	T	U	V
X	X	Y	Z	A	B	C	D	E	F	G	H	I	J	K	L	M	N	O	P	Q	R	S	T	U	V	W
Y	Y	Z	A	B	C	D	E	F	G	H	I	J	K	L	M	N	O	P	Q	R	S	T	U	V	W	X
Z	Z	A	B	C	D	E	F	G	H	I	J	K	L	M	N	O	P	Q	R	S	T	U	V	W	X	Y

Figure 11.6 The Vigenere Table

referred to Figure 11.6. When it comes to the vigenere table, it consists of 26 rows and 26 columns which are filled with the alphabets 26 times in a looping format. That is, cyclically each alphabet is shifted to the left from the previous row alphabets.

Encryption In the encryption process, from any of the rows, different alphabets are used for generating the cipher text.

Decryption In the decryption process, the vice versa operation like encryption will be performed. But the only difference is that the encryption operation will be done on the plain text and the decryption operation will be done on the cipher text generated to obtain the plain text.

MD5 Hashing Techniques A hashing technique accepts a message input of any length and provides the message output of a fixed length, which is also said as digest value. This digest value will be used for plain text message authentication [8–10]. For providing a secured cryptography hashing algorithm and for digital signatures

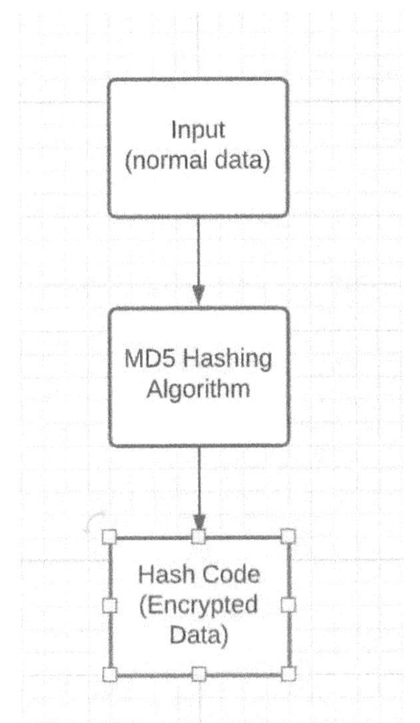

Figure 11.7 MD5 Hash Technique Working

authentication, this MD5 hash technique is designed. This helps in verifying data integration and avoids data corruption. As discussed earlier, this MD5 hash functions, as shown in Figure 11.7, an input message of any length and generates a fixed length of 128-bit output, which is also known as the message digest value of the input message. It was mainly designed for digital signatures authentication.

If some large files are highly confidential, then it must be compressed in a cryptography secured manner before starting the encryption process with a secret key of a particular user using cryptography algorithms. This states the digital signature of the files. So, for providing those digital signatures, this MD5 hash technique will be very much useful.

Drone Frame Using the Tinker cad tool, a basic drone has been designed and all modules get embedded within the drone, as shown in Figure 11.8.

All the components required for the design of the drone along with the processing unit such as Arduino is configured using the Tinker cad tool and expected output of the design can be visualized

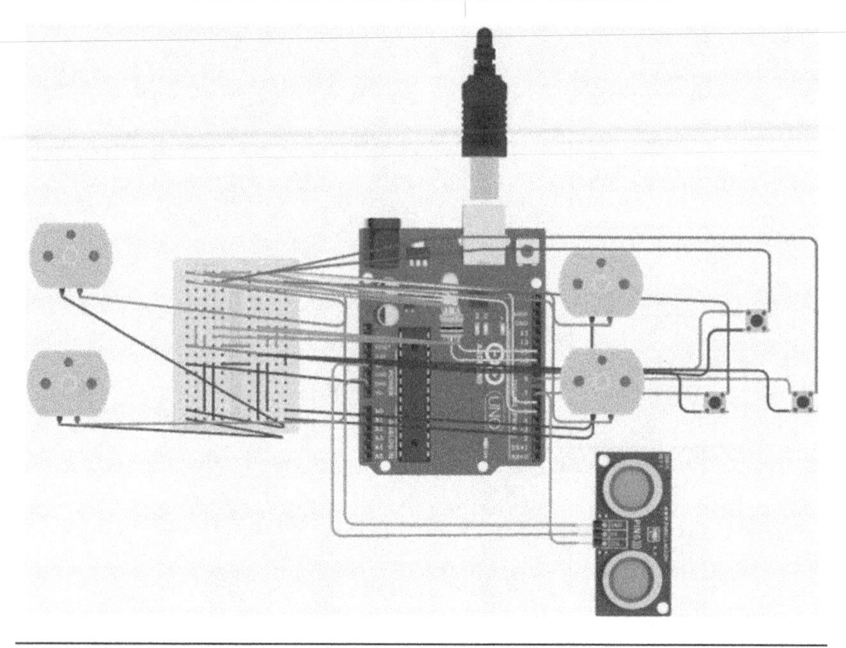

Figure 11.8 Design of Drone Simulation

by simulating the design in the Tinker cad tool. This may help us to debug the circuit if the expected output gets varied from the observed output.

Implementation

This section focuses on the implementation methodologies in perception with neural network models, such as convolution and pooling techniques, along with image classification implementation technique, such as VGG16. Image cleaning to its preprocessing and image visualization using RGB and HSV color models are also discussed here.

Implementation of Methodologies of the Neural Network Models

Figure 11.9 signifies the implementation methodologies using neural network containing convolution and max pooling techniques.

Convolution: Convolutions are a type of operation that can be used to learn representations from images. They include learnable kernel slides over the image performing element-wise multiplication with the

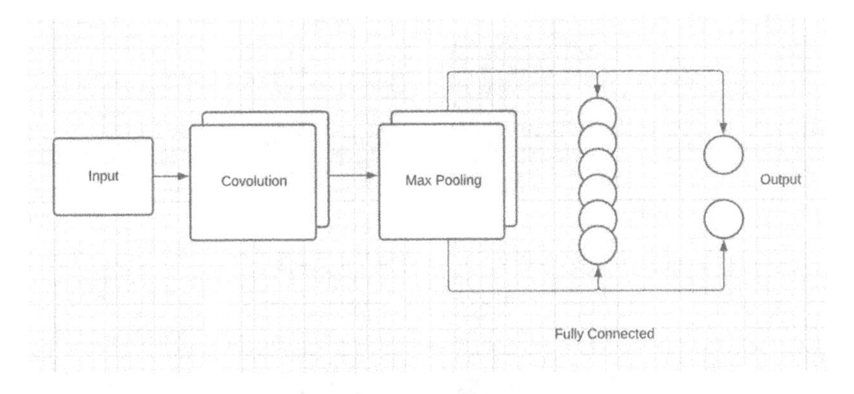

Figure 11.9 Implementation of Neural Network Models

input. The specification permits for parameter sharing and translation in-variance. Given below is a continuously updated list of convolutions.

Pooling: Max pooling is an operation that calculates the utmost value for patches of a feature map and uses it to form and create a down-sampled feature map. It adds a little quantity of translation in-variance – that means translating the image by a small quantity does not considerably affect the values of most pooled outputs.

Implementing VGG16

In this model, VGG16 is implemented for image classification. The following steps are involved in image classification process.

Import Packages In the first step, all the necessary packages for image classification using VGG16 are imported from Keras library in Jupyter Notebook with Python programming language.

Import DataSet Here, the dataset containing psychoactive drug plants images should be imported into the Jupyter Notebook for further data preprocessing of the dataset. The dataset consists of info of 12 plant species. Also, the sample images from the dataset will be displayed to check whether the data have been successfully imported into the notebook. Figure 11.10 shows the sample visualization of crop seedling images from the imported dataset.

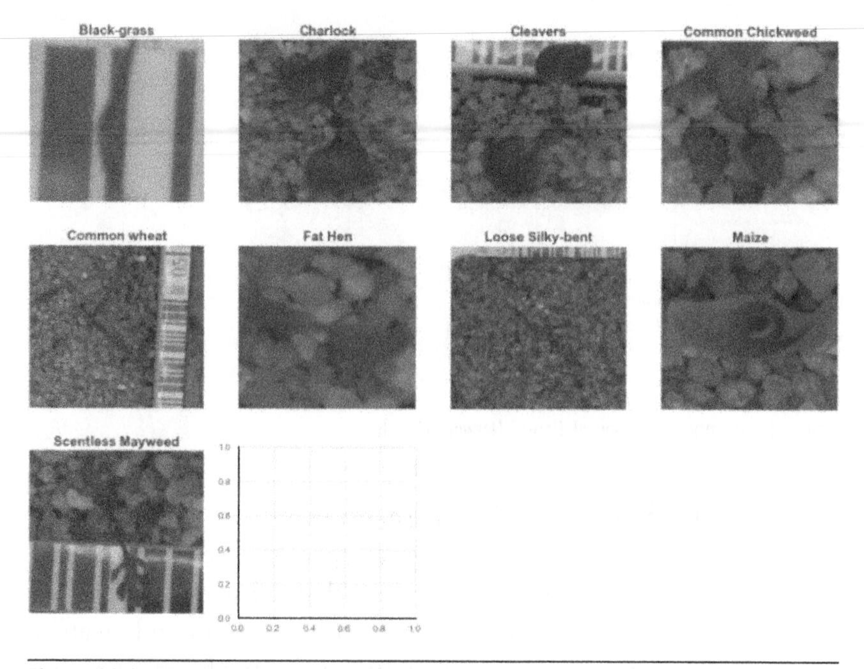

Figure 11.10 Visualization of Crop Seedling Images

Image Preprocessing The images in the dataset will go under preprocessing. In the imported data while importing it into the notebook, there may be a chance of missing values and null values. The dataset may also contain the noise data in it. For retrieving those missed values and avoiding noise data, data preprocessing will be done.

Import Target Image Input Processing the image to get rid of the unnecessary pixels is depicted through Figure 11.11. Feeding this picture to the model directly will result in a target leakage. The classifier will learn the stones and the container's pixels and build its prediction upon these pixels as well. This is something that will be avoided. Preprocessing starts with by doing object segmentation from the input image which is based on the RGB colors using a Python library called OpenCV, as it is the most popular computer vision-based library for manipulating colors in an easy way.

Visualize the Plant in RGB Visualizing the small-flowered Cranesbill image we opened above in RGB space to see the distributions of the

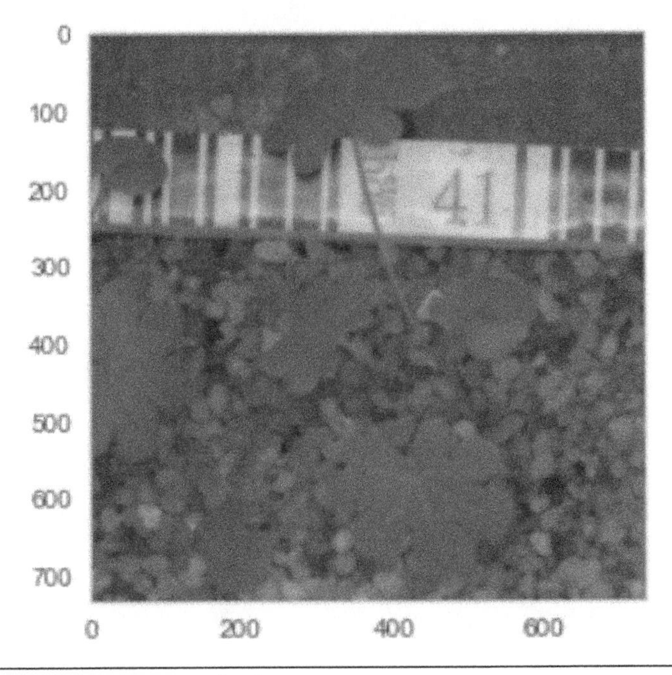

Figure 11.11. Target Image as Input

color pixels. RGB is considered an "additive" color space, and colors are visualized in red, green, and blue with the background color black.

From the given RGB visualization in Figure 11.12, green parts of the image are mixed with red and blue values, segmenting our plant image out in RGB space based on the ranges of RGB values. The solution is to try another color space, and we go next with HSV color space.

Visualize the Plant in HSV HSV is the alternate approach for RGB color model. HSV stands for hue, saturation, and lightness value, and represents the mixing of different colors with saturation dimension resembling. It defines a threshold value to help it become a binary image with the color, as mentioned in Figure 11.13.

In HSV space, our green plant are much more localized and visually separable. The saturation and value of the greens do vary, but they are mostly located within a small range along the hue axis. This is the key point that can be leveraged for segmentation.

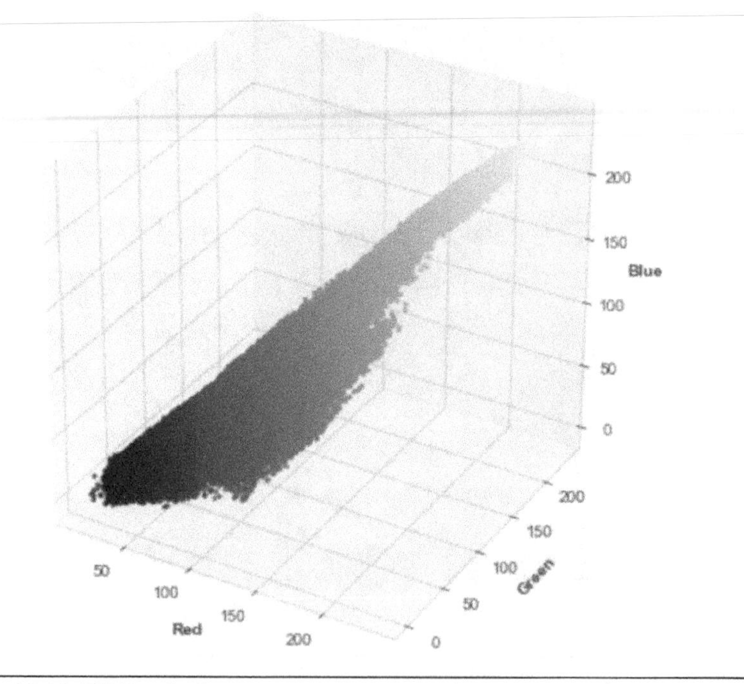

Figure 11.12 RGB Visualization

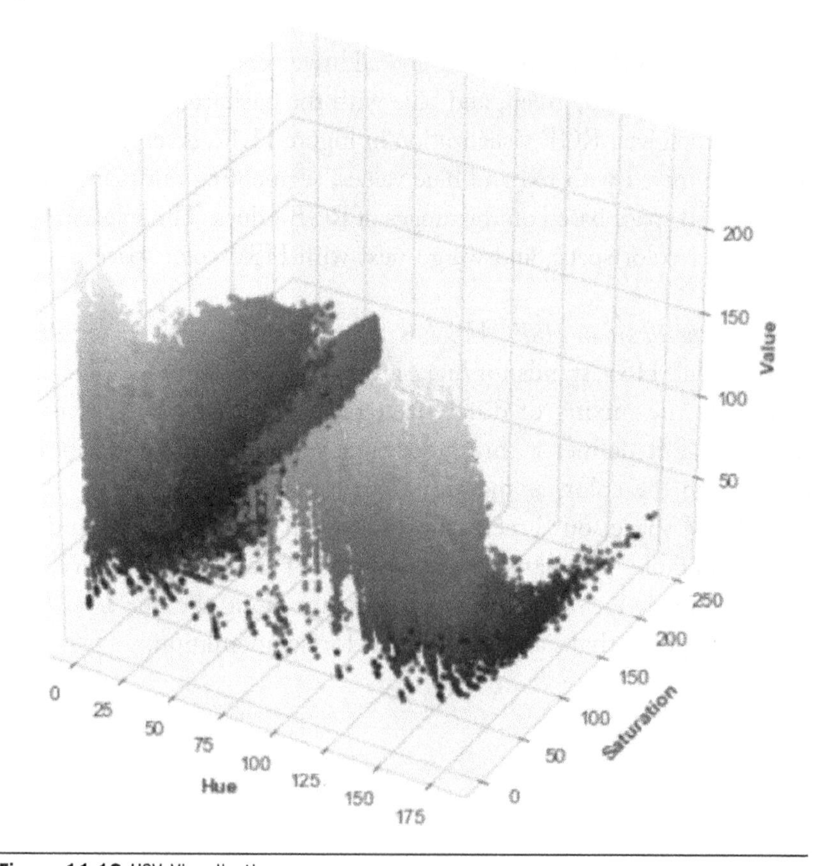

Figure 11.13 HSV Visualization

Segmentation

Color-Based Segmentation To pick out a range of color, we can use a color picking app online. I found the green color ranges in a forum:

- Minimum green(H=36, S=25, V=25)
- Maximum green(H=70, S=255,V=255)

The OpenCV function cv2.inRange() is used to ensure the threshold values of the plant image ranges with the minimum and maximum green values.

InRange() takes three parameters: the image, the lower range, and the higher range. It returns a binary mask (and array of 1s and 0s) of the size of the image where values of 1 indicate values within the range and zero values indicate values outside [11] (Figure 11.14).

- colormin=(36, 25, 25)
- colormax=(70, 255,255)
- plot_mask(p1, coloring, color max)

The result of our initial threshold is not bad, since some pixels of the plants were missed. Later, it has been manually adjusted with saturation and brightness values to get a better mask [12] (Figure 11.15).

- new_colormin=(25,50,50)
- new_colormax=(80,255,255)
- plot_mask(p1, new_colormin, new_colormax)

Segmentation Function Now, a function that applies segmentation on the rest of the images is created. This function will be used to feed

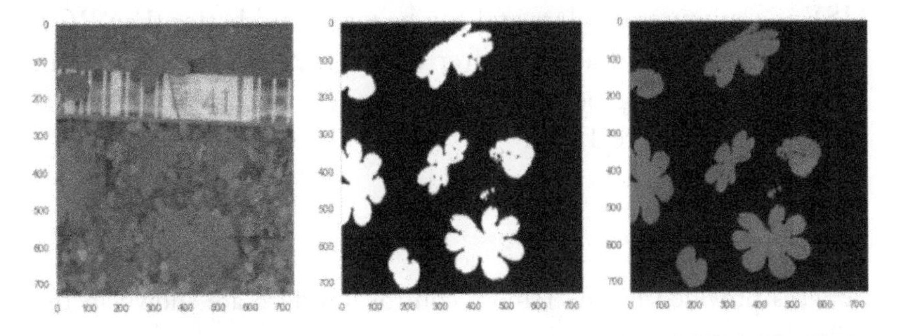

Figure 11.14 Segmentation of the Input Image

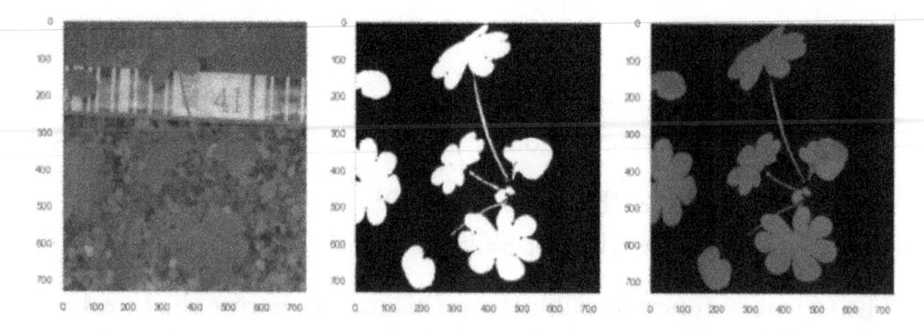

Figure 11.15 Segmentation of Image without Missing Values

the computer vision model with the segmented images instead of the original ones.

Computer Vision Model

This shows plant images to a pre-trained VGG16 convolutional neural network. It is trained with both original and segmented images to find out the impact of the target leakage on training. It involves VGG16 with original plant image.

Experimental Results

Training Images

Trained models get an input of the collected images and the model performs the segmentation for the target image and then the image gets converted into the RGB color model.

Once the RGB color model is not clear, the images are converted and represented in the HSV color model. In this model, HSV is a mixture of colors and can be visualized better than RGB model. Images get classified after the segmentation, as shown in Figure 11.16. After performing the model, the predicted results are shown in Figure 11.17.

Remote Interface Screenshots

Services provided by the remote control interface are shown in the following screen shots.

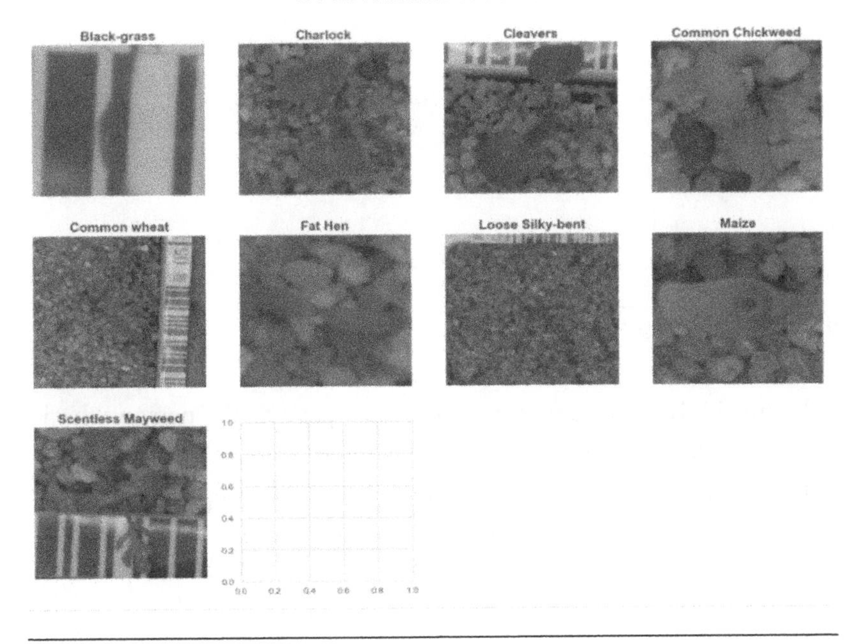

Figure 11.16 Training Images

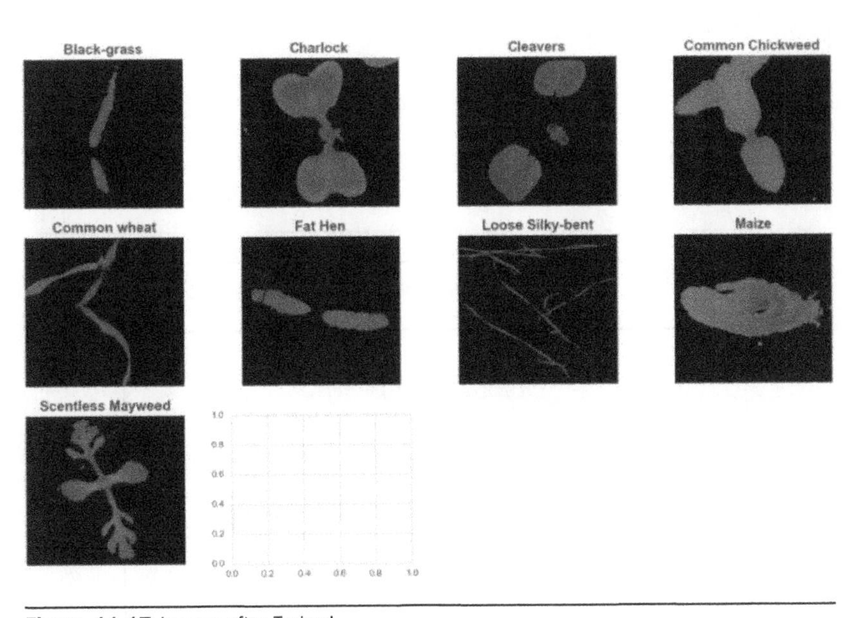

Figure 11.17 Images after Trained

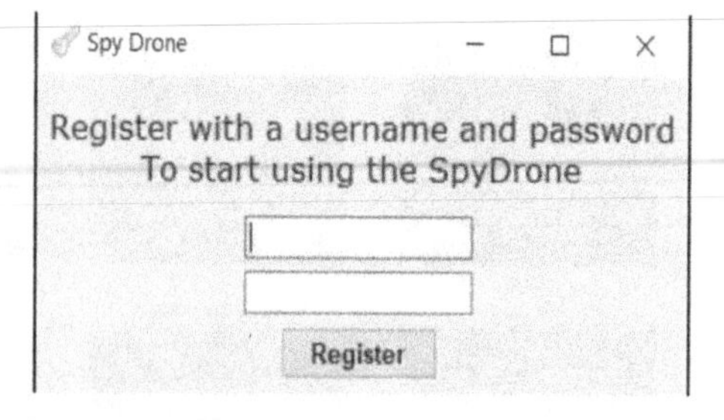

Figure 11.18 GUI Registration Page

Figure 11.18 represents the registration page to the person who is going to access the drone.

Figure 11.19 depicts the GUI for login into the application where the person has to register to login into this module.

Figure 11.20 contains three options whether to add the location, see the details in the list, or search for any data in the list.

Figure 11.21 contains the details of the authorizer who controls the drone and location details which are in encrypted format and all text fields must be filled.

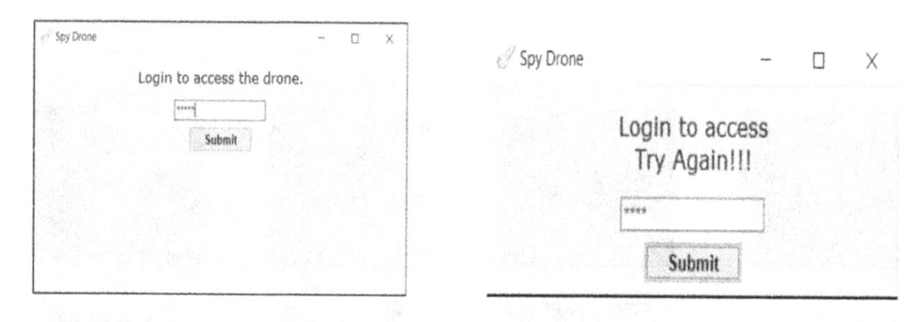

Figure 11.19 GUI Login Page

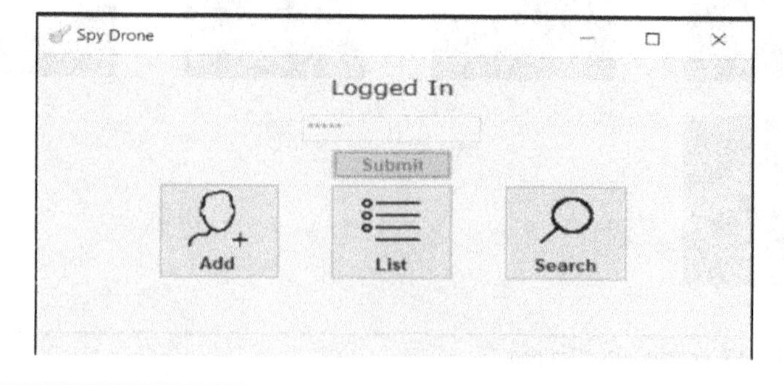

Figure 11.20 Home Page of GUI

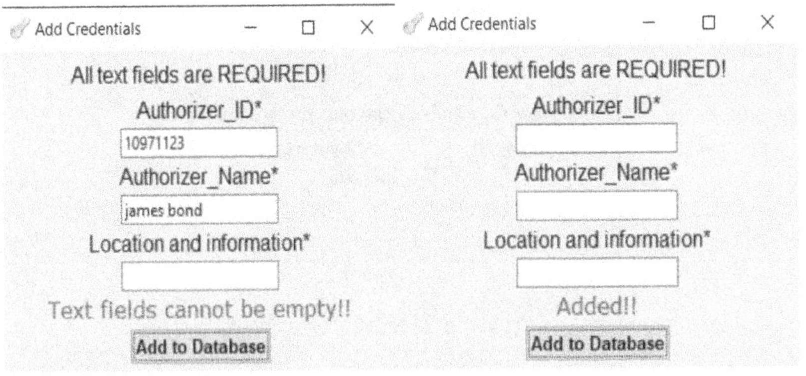

Figure 11.21 Adding Collected Information

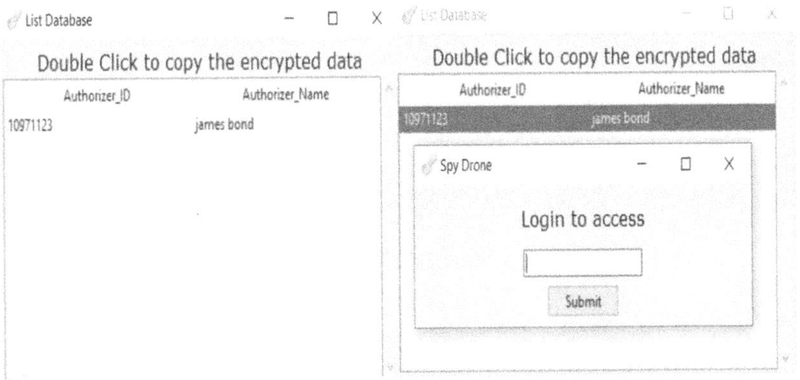

Figure 11.22 Added Details

In Figure 11.22, added details are viewed but the location and details of the identified plants are not viewed. Only the person gets authenticated and then the information is viewed.

Figure 11.23 is used for searching the details of the authorizer.

The security for this GUI application will be provided using cryptographic algorithms like the vigenere cipher algorithm and MD5 hash techniques. One of the most important things is that the drone will be incorporated with the shortest path algorithm for sending the drone to different specified locations and for returning to the controller quickly. Figure 11.24 represents the entered location saved in an encrypted format.

Search
Search
Double Click to copy the encrypted data
Authorizer_ID
Authorizer_Name
10971123
james bond

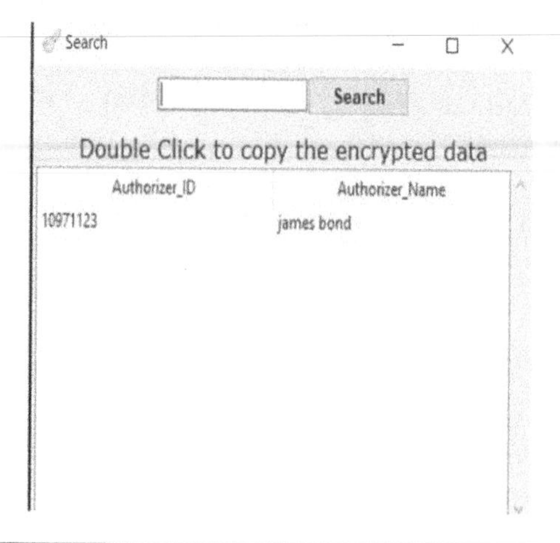

Figure 11.23 Search Module

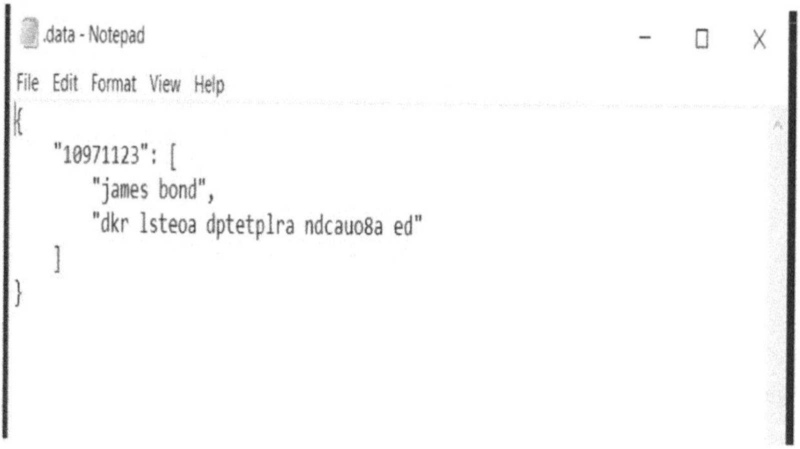

.data - Notepad
File Edit Format View Help
{
"10971123": [
"james bond",
"dkr lsteoa dptetplra ndcauo8a ed"
]
}

Figure 11.24 Encrypted Data

Conclusion

In this work on "IoT-Automated Spy Drone to Detect and Alert Illegal Drug Plants for Law Enforcement," mainly three modules that have been discussed. The first module is that there will be an autopilot that is integrated with a camera, in-built GPS navigation, and the image classification application for classifying the psychoactive drugs from a group of plants being grown in the specified locations. The second module is the image classification application, in detail. This application will be used for classifying the plant species that are captured by the drone using a camera. For building this image classification application, a convolutional-based neural network called VGG16 neural network model has been used for classifying the psychoactive drugs from the group of plant images. This VGG16 neural network model has been implemented in the python programming language using the Jupyter Notebook as a platform from the Keras libraries. The internal processing of image classification has been discussed in this work. Finally, the third module is about the remote control interface. There should be a remote controller that is controlled by the controller to control the drone. The unique feature of this remote controller is that there will be an inbuilt GUI application in it for the controller to store the collected information from the drone. The collected information and user details must be secured to avoid data leakage. Here, comes the cryptography algorithms used for encrypting the gathered information and also for providing confidentiality, authentication, and integrity for the application. The controller after collecting the information will store the details of it in this GUI application, which is associated with a database, so that he can retrieve the information in the future and take proper actions on the identified landowners. The security for this GUI application will be provided using cryptographic algorithms like the Vigenere cipher algorithm and MD5 hash techniques. One of the most important things is that the drone will be incorporated with the shortest path algorithm for sending the drone to different specified locations and for returning to the controller quickly. This helps the drone to select the best and the shortest path among all the possible paths to reduce energy and time consumption. The output screenshots of the image classification and the GUI application have been given in the experiment results section. These are all the things that are discussed in this work. Finally, to conclude that the main goal of this Spy Drone application is detecting the psychoactive drugs in a specific location and collecting the information that is required for taking further actions using a drone. This helps the government sector people to control the illegal activities of drug usage and production efficiently without using more manpower.

References

[1] Aized Amin Soofi, et al., "An Enhanced Vigenere Cipher for Data Security," *International Journal of Scientific & Technology Research,* vol. 5, no. 03, pp. 141–145, 2016.

[2] Z.-p. Wang et al., "Single-Intensity-Recording Optical Encryption Technique Based on Phase Retrieval Algorithm and QR code," *Optics Communications,* vol. 332, pp. 36–41, 2014.

[3] Shweta Mishra et al., Hashing Algorithm: MD5, Department of Computer Science & Engineering, Echelon Institute of Technology, Faridabad, India; JB Knowledge Park, Faridabad, India, 2013.

[4] Piotr Kardasz and Jacek Doskocz. "Drones and Possibilities of Their Using". *Journal of Civil & Environmental Engineering,* vol. 6, Piłsudskiego, Wrocław, Poland, 2016.

[5] VGG16 and VGG 19 functions are referred from following the link https://keras.io/api/applications/vgg/

[6] Karen Simonyan and Andrew Zisserman, "Very Deep Convolutional Networks for Large-Scale Image Recognition," In: *International Conference on Learning Representations,* 2015.

[7] Ayon Dey, "Machine Learning Algorithms: A Review," *International Journal of Computer Science and Information Technologies,* vol. 7, no. 3, 2016, 1174–1179.

[8] Working of VGG16 model is referred from following the link https://www.geeksforgeeks.org/vgg-16-cnn-model/

[9] Algorithm of MD5 hashing techniques are referred from following the link http://practicalcryptography.com/hashes/md5-hash/

[10] Mary Cindy, Ah Kioon, Zhao Shun Wang, and Shubra Deb Das, "Security Analysis of MD5 algorithm in Password Storage," *Applied Mechanics and Materials,* vol. 347–350, pp. 2706–2711, 2017.

[11] Vigenere Cipher, Eric Conrad, and Joshua Feldman, *CISSP Study Guide* (3rd Edition), 2016.

[12] Jean-Paul Yaacoub, Hassan Noura, Ola Salman, and Ali Chehab, "Security Analysis of Drones Systems: Attacks, Limitations, and Recommendations," *Internet of Things,* vol. 11, May 2020.

12

Expounding k-Means-Inspired Network Partitioning Algorithm for SDN Controller Placement

J. PUSHPA[1]
AND PETHURU RAJ CHELLIAH[2]

[1]*Jain University, Bangalore*
[2]*Site Reliability Engineering (SRE) Division, Reliance JioInfocomm. Ltd., Bangalore, India*

Contents

DOI: 10.1201/9781003119838-12

Introduction

The power of Artificial Intelligence (AI) significantly brought the new paradigm to resolve many challenges which were heaped with inefficient solution. AI adopted many robust and sophisticated algorithm and also the analysis of AI infuses in for many scenario which gives accurate and efficient resultant values. In this chapter will discuss about the one such algorithm that is k-means algorithm which can be amalgamated by any technology for different kinds of problems.

Before we discuss in detail about the k-means algorithm, will discuss some of the scenario which direct to choose the kind of algorithm/approach for analyzing/categorizing/resolving the problems.

In the field of electing leader such as master node in networking, ranking the quality or grading the system are all not continuous or common value. Those can be consider as for discrete data. Discrete samples are countable/measurable which require linear regression techniques. On the other hand if value have similarities such as common symptom, same type of traffic, similar pattern or recursive samples than it is continuous data. For continuous data which have similarities which are not countable. For continuous samples which are massive in volume with unlabeled data need to group with non linear counter part by clustering. Clustering the data is one of the prominent way to classify or group the values in which most of clustering algorithm adopt the euclidean/correlation distance formula for segregation non linear values.

If we look the idea behind clustering algorithm is basically originated for analysis purpose and its journey is from past many decades. Our main objective is to discuss one of the traditional and efficient clustering algorithm that is k-means clustering algorithm. Clustering based problem are in all the fields from genetic related analysis to computer network traffic analysis. Most of the researcher have opted k-means algorithm for clustering. In this chapter will discuss on k-means role in advance technologies, properties and its efficiency.

k-means algorithm also know as Llody's algorithm which is much influenced on vector quantization. Basically, this algorithm work as all small K similar group moving towards centroid with minimal distance and computing the distance is computed using Euclidean

Distance formula. According to [1], k-means algorithm is the process for partitioning N-dimensional population into K sets. Many definition or interpretation of k-means can be find in various field. J. Andrew [2,3] has worked on augmented supervised and non supervised clustering with k-means algorithm which has resultant to the better convergence and classification. Integration of k-means algorithm with other algorithm like LVQ algorithm [4] has increased the performance of clustering.

k-means can not only integrate with different methodology but also applied in the field of Genetics, Machine learning, Artificial intelligence and many more. Hence it can be declared as versatile algorithm for clustering problems. Section 2 will give an origin and hierarchical of k-means to which it belongs in machine learning. In Section 3, we discuss the k-means algorithm and its working structure along with its prime properties. k-means is applied in many technology which conjugate with other methodologies for better efficiency, this discussion helps to route our exploration for solving the problems.

And in last two section the discussion extend on the implementation and applications in which k-means plays an vital role.

Leveraging on Unsupervised Learning

Its a greater part to know new technologies and their application basically as applied science to experience the benefit out of it. Learning Data Science, Artificial Intelligence and Machine Learning make the technology work in am very robust and acute way. There are a plethora of use case which helps to understand in deeply about the algorithms and approaches which helps to make the world connected through IoT devices. In this section we learn more about the Unsupervised learning and its importance.

IT is enriching in all the domain mainly fascinated to develop IoT products which is widely in high demand. The technology behind IoT is not only Arduino, Cloud Technology but also the intelligent algorithm of machine learning which adopt either regression, classification or clustering algorithm.

These algorithms are categorized into Supervised, Semi Supervised and Unsupervised learning, it basically depends on the type of the data fed to the system.

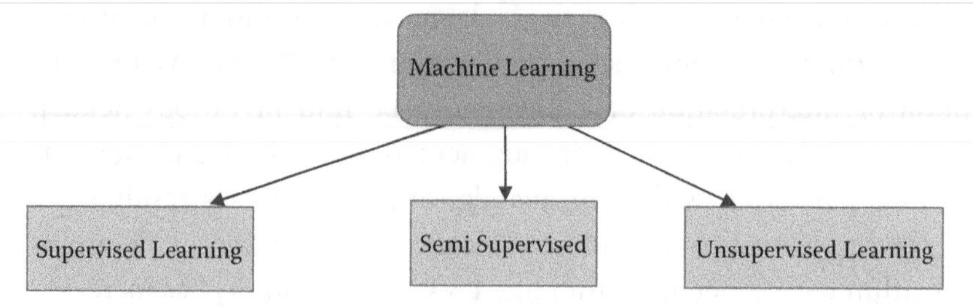

Considering an automated system such as Face detection, Weather forecasting, Stock Price prediction which are operating on the trained data or predicting based on the previous knowledge in these cases few information is already stored and based on it the output will be predicted such kind of mechanism called Supervised learning. In supervised learning if some situation arises like malicious traffic or unknown symptom in the patient or any unusual load then the prediction may fail to give the accurate output this can be overcome by regress evaluation process with targeted result which has no predefined label termed as Unsupervised learning. As discussed in the previous section k-means is the most popular algorithm which also comes under unsupervised learning due to it evaluation process and the rate of convergence.

Before understanding in detail about the k-means algorithm of unsupervised learning, will understand the basic characteristic of unsupervised learning.

Unsupervised Learning is a technique to unveil the hidden pattern or structure in the input data set. As shown in the Figure 12.1, unsupervised learning methodology is classified into clustering, dimensional reduction and association.

Let us consider an example to get the difference between those methodologies, as the internet connectivity is getting better the malicious attack is the biggest challenge in networking. To overcome this challenge the network traffic should be classified and identifies the DDoS attack.

In this scenario not only the classification algorithm alone can be used along with that data size should also matters because the network traffic is collection of enormous packets with varying size and also need to establish the relation to study the association to capture the similarity.

Classification → **Clustering**
Data Size Reduction → **Dimensional Reduction**
Similarity/Relation → **Association**

I. *Clustering:* Data are been partitioned into distinct subset based on the parameter such as distance from the centroid.
II. *Dimensional Reduction:* Reducing the raw data set into manageable sized data set with integrity is the task of Dimensional Reduction.
III. *Association:* Is a mechanism to identify the rule for establishing the relation.

Relating the unknown value in a large data set is a complex process, as depicted in the Figure 12.1. Classifying the large data set using the mention methodology takes an deep analysis in which k-means is categorized as clustering algorithm. But, some of the data set dimension value will be more in number which require mechanism to simply and associate it based on the relativity.

Principal Components Analysis(PCA) play an important role in segmentation to reduce the dimensional features. Assume if the data contains more than 20 features, then clustering perhaps results to an inefficient cluster. Hence, standardizing and reducing the low impact features are needed. Further, before proceeding with k-means clustering it is necessary to compute the PCA.

An unsupervised learning is a wide topic which is applied in most of the technology, we are focusing on clustering methodology and also discussing on most popular clustering algorithm which help us to understand and differentiate with k-means algorithm.

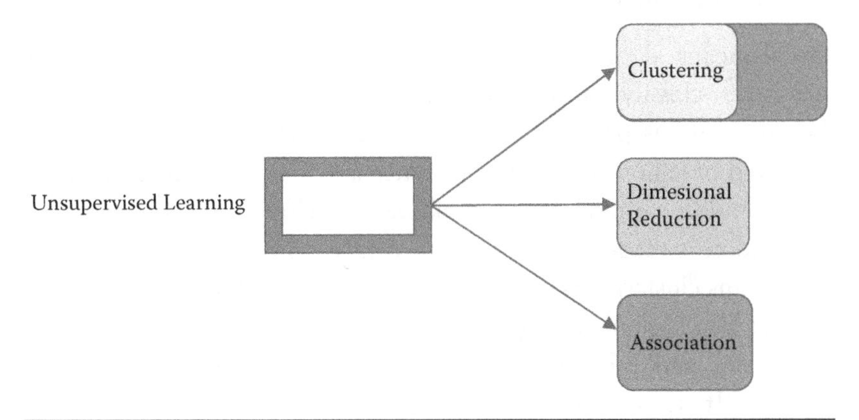

Figure 12.1 Methodology of Unsupervised Learning

Table 12.1 Clustering Algorithms

TYPE	TASK	EXAMPLE
k-means, K-Mediods	Clustering the data points by calculating the centroid distance.	Compression, Data Mining.
Hierarchical Clustering	Nestested clustering with levels	Anomaly detection, Deep Learning.
Density-Based Clustering	Clustering based on density in which the grouping will done based on the quantity.	Text clustering, Image clustering, Outlier detection.
Spectral Clustering	Clustering based on graph.	Community detection, image segmentation, network.
Gaussian Mixture Models	Clustering based on the probability distribution for a given data set.	Speech recognition, Signal mapping and positioning
Nearest Neighbors	Grouping the data points which are close using exhaustive search.	Machine Learning
Hidden Markov Models	Markov model for stochastic process based on current state	Stock Market,Position of Chromosome in genes

Clustering is a methodology which classify based on the requirement basically to analyze the pattern or organize the data else to recognize the pattern. For implementing this clustering many strategic algorithm are discovered by researcher in which k-means is the most popular clustering algorithm. As discussed in [5–7], the algorithms comes under clustering is tabulated below in Table 12.1.

Clustering

k-Means Algorithm

The k-means algorithm is introduced by J. Macqueen [8] for the method to classify and analyze the asymptotic behavior of data set, the proposal has proved the optimal classification of sets which are measured based on the relative distance and also concluded that as k value should be minimal which will be than close to mean value, this observation is made on multivariate with varying dimensional value.

k-means clustering is a partitioning clustering which partitions the data set into k number of group which contain the equal size in both the cluster and the objects of one cluster is not belong to other clusters. It uses Euclidean/Squared Euclidean/Manhattan/cosine distance algorithm for the distance between the centroid and the other nearing objects. This process will continue in an iterative way to group all the objects closer to centroid, hence the objects can not be in a single cluster till the final convergence.

Finding the k-means is depicted in the diagram Figure 12.1 given in Figure 12.2.

As presented in flow chart, the flows are discussed below with few question.

STEP 1: Indicate the considering the data set Ss = {x1, x2, x3, ….xn}:
In this step few things need to consider such as:
What kind of data set?

- Unsupervised data set is more suitable for k-means clustering
What is the limitation of data set?
- Big O(better performance) for small data set and number of iteration increases for large data set which will non-optimal solution.

STEP 2: Initialize the value of K.:
To determine the optimal number of cluster there is no definite rule to choose it. As discussed in datanovia [9], there are two startergies

a. direct method
b. statistical testing method, (will discuss it in detail in the next section)

STEP 3: Assign the Centroid to each cluster, basically euclidean distance formula is applied to bring the data point closer to the centroid for grouping. Centroid C = {c1, c2, c3, … ck}:
Euclidean Distance =arg min $dist(ci,x)^{2}$', where ci belongs to C.

STEP 4: Compute the new centroid which will be mean value in a group, so new centroid will be emerged and regrouping may take place.

$$Ci = 1/|Si|\Sigma xi, \quad xi€S.$$

If reposition of centroid then repeat then step 3.:
else:
goto step 5

STEP 5: Once the cluster is static, k-means clustering is said to be converged.:
The above steps not only explain the process in k-means but also some of the parameter such as k value and distance formula between the data points. Clustering algorithm not only aims to group the data points but also need to consider some of the

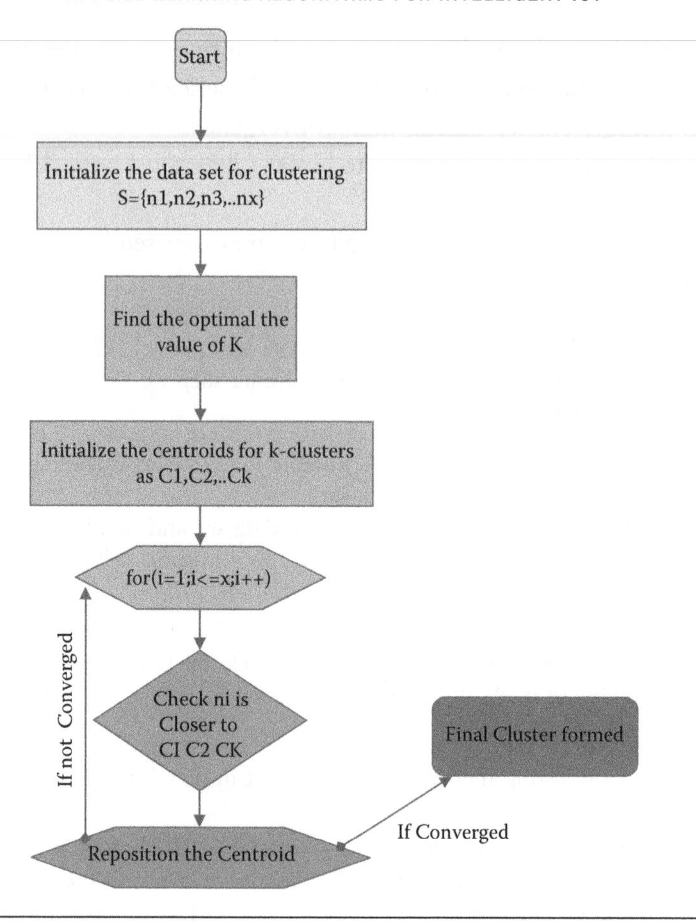

Figure 12.2 k-Means Algorithm Flow Chart

coefficient factors such as i*deal distance value between the data points in group, the distance between the group, optimal number of cluster and number of data point in a cluster*. Will discuss those coefficient factor as the properties of k-means algorithm and also see some the example on it.: Many of the researcher have used the k-means to resolve their problem for the certain kind of data set which kind of Data set can use k-means algorithm is our next discussion.

Data-Set for k-Means Algorithm

- It is important to understand the pattern of data before applying any algorithm. k-means gives better convergence when data set is in *spherical / convex*shaped structure and failed for non-convex shaped.

- k-means clustering algorithm fails to give good convergence when the data set contains outliers, the density spread of data points across the data space is different and the data points follow non-convex shapes.
- The K-means algorithm needs the value of K which determine the number of cluster which has to entered manually.
- One of the feature scaling ensures that all the features get same weight in the clustering analysis.

Distance Formula of Centroid in k-Means Algorithm

All the data points of a cluster in k-means algorithm is closer to the centroid of that cluster, and the distance between centroid is computed using Euclidean distance formula and there are other alternative formula also been used based on varying requirements as shown in the Figure 12.2.

The k-means algorithm is an solution for many problems which helps to reveal the hidden pattern, we learnt how the iterative process helps to create the clusters in which two factors are playing an prominent role one is distance between the centroid and the other one is the value of k. In the Figure 12.3 various distance formula can be associated with k-means for variety of problems.

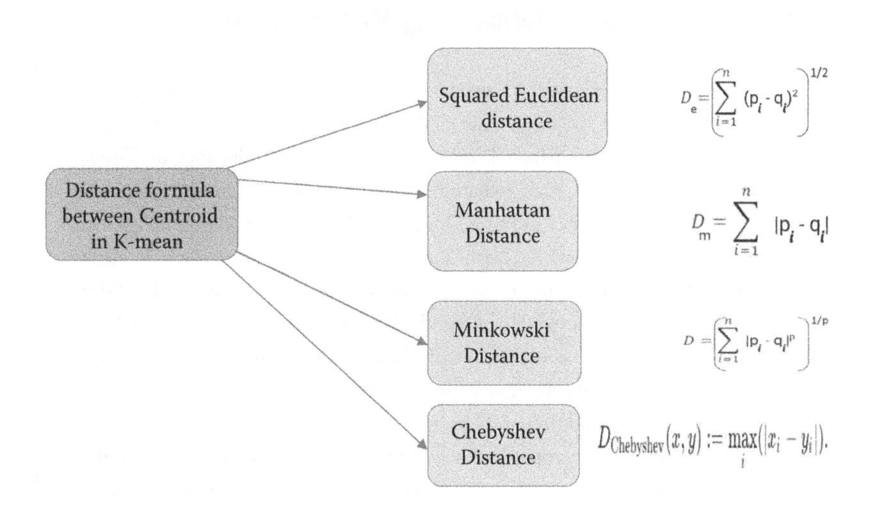

Figure 12.3 Distance Formula

Let us consider few sample data which helps to understand the applicability of those methods and direct to choose the appropriate distance formula.

Case 1 k-means using Squared Euclidean distance formula.

Squared Euclidean distance formula is basically used to minimize the distance within in the cluster. Jenks has come also used the concept of k-means using squared deviation mean for cartography [10], and the squared distance is used in many scenario's.

k-means has proved over many complicated algorithm with a better performance in classification, as discussed in [11] k-means is suggested to recover the mixture of matrix which have more sparsity and also unlock the ICA- like properties.

k-means is also gives effective result to classify the network traffic, as described in [12] its been proved as effective and faster model to classify the network traffic. Data set used in [12] by Erman J are two empirical packet traces, One is a publicly available packet trace and other one is full packet trace such as Auckland IV 2 and the University of Calgary respectively. The classification of internet traffic to clustering using DBSCAN,k-means and Autoclass in which the building time of model is comparitively less and more efficeint in k-means whereas Autoclass is very time consuming and DBSCAN accuracy is very low. This conclude that k-means is faster, accurate and efficient for clustering.

Case 2 k-means Clustering using Manhattan Distance.

Data mining is widely used in many field for extracting the data in high dimensional data set at the same time challenges of searching and retrieving the relevant data also triggering task. k-means with Manhattan distance or Taxicab formula is well suited in such situation. As discussed in "On the Surprising Behavior of Distance Metrics in High Dimensional Space" by Charu C. Aggarwal [13], high dimensional data set with 2 or 3 D spatial system which is having fractional matrix is very common in data mining in which applying Euclidean distance which is best suited for integral distance is irrelevant for this hence Manhattan distance has come out with best suitable and efficient distance formula. Dibya in [14] has discussed the performance of Manhattan distance on data set such as Iris and Wine Data set and the resultant proved that the computation time is less compare with Euclidean and Cosine distance formula.

Hence Manhattan distance is more suitable for the data set which are in the grid format between a pair of coordinator. Though the computation is expensive it is significantly resistant.

Case 3 k-means Clustering using Minkowski Distance.

Minkowski distance algorithm is used to measure the distance between two points in a normal vector space as a generalization for both Manhattan and Euclidean distance. It is an effective in fuzzy logic, networking, spatial data and many. In [15], the performance of *minkowski distance* is proven on spatial distance for measuring the distance on the basis of time.

Minkowski Distance has not only used for calculating the distance but the convergence result with reduction of error and increased the accuracy with less computation time as discussed in the field of image segmentation in [16].

Many researcher has used this distance formula and also extended their work for improvement of this algorithm which is delimited for the value of p=1 and 2. As we can see in "Fizzy clustering using Minkowski Distance" in [17] and resultant with efficient and well applied to empirical applications.

Case 4 k-means Clustering using Chebyshev Distance.

k-means clustering is versatile due its performance with varying distance formula, as discussed above we extended our discussion about chebyshev distance formula. It gives the maximum absolute distance. Major applications are Chessboard and warehouse logistics.

Chebyshev distance is used in many field such as in machine learning, IoT devices and AI as this gives he distance of an object moving away or with maximum distance.

Below are the some pattern of data representation for the above discussed distance formula which help us to identifies the pattern before applying the distance formula.

In this chapter, we have discussed four main distance formulae with respect to k-mean which represents the qualitative behavior of the different distance metrics for measuring proximity in high

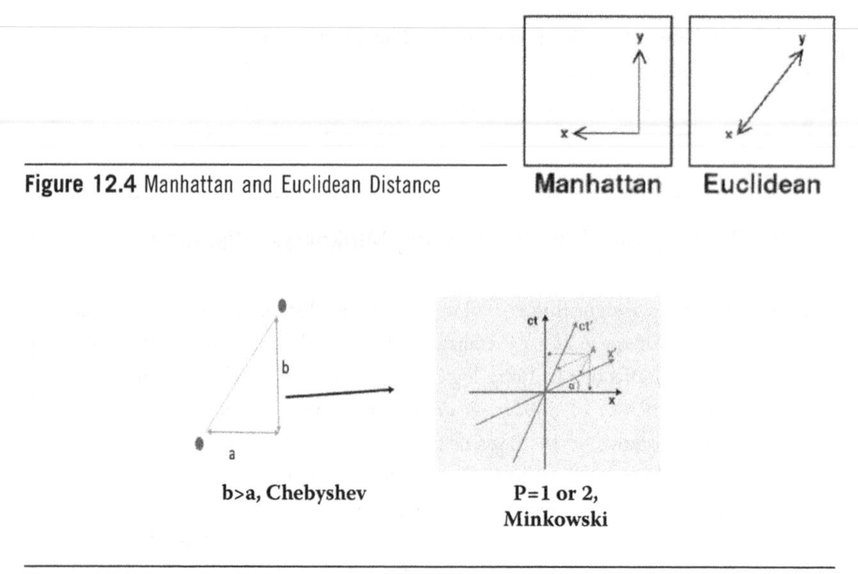

Figure 12.4 Manhattan and Euclidean Distance

Figure 12.5 Chebyshev Distance and Minkowski Distance

dimensions. The particular choice of distance metric which is used from problems such as clustering, categorization, and similarity search which depends on some notion of proximity. Every formula have its own significant characteristic for the specific pattern (Figures 12.4 and 12.5).

In the next section will discuss on mechanism to determine the optimal number of k in k-means algorithm.

Mechanism to Compute the Value of K

In k-means algorithm k indicate the number of clusters, by default k is been taken as random number which make be not appropriate and significantly no valid point in it to over come that many indices are used by researcher. Here, mainly two point to be noted one is about the value of and another is to find whether in the data set is there any possibility to form cluster or not which is decided based on null hypothesis and alternative hypothesis [18], we are not covering in detail about hypothesis in this chapter.

Before applying those method to find the value of K, identify the dimentionality of data set because euclidean distance is not the

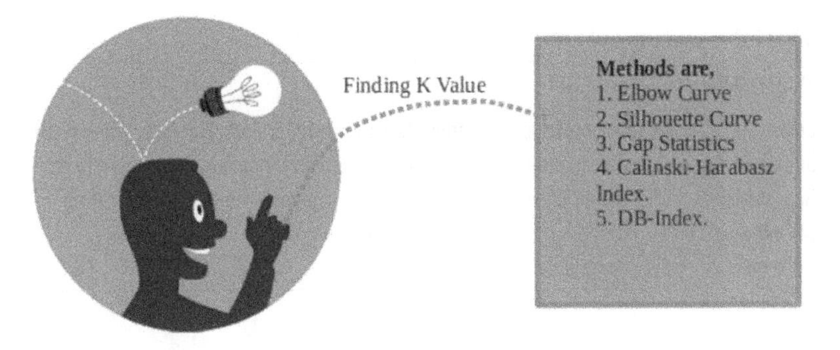

Figure 12.6 Methods to Find K

suitable solution for multidimensional data set. Hence PCA should be implemented on data to speed up the computation.

Determining the k-values have many method in which five main popular methods are discussed below and represented in the Figure 12.6 in which most of the methods are dependent on Within Sum of Square (WSS) and Between Sum of Square (BSS) [19–21].

$$WSS = \sum \sum_{x \in Ci} (x - ci)^2$$

$$BSS = \sum_i |C_i|(c - ci)^2$$

1. **Elbow Curve:** This method choose a range of candidate values of k, then apply k-means clustering using each of the values of k to draw the inertia. Represent the average distance of each point to its centroid of the respective cluster. The value of inertia decrease as k increases and stop declining after reaching saturation value of K.
2. **Silhouette Curve:** It finds all the average distance within a cluster and their distance with other cluster object. Once it computes the minimum value will be taken. It is all about the relation between cohesion and resolution.
3. **Gap Statistics:** In this methodology the value of within sum of square cluster is compared with the different value of K.
4. **4.CH-Index:** The CH-index is used to find the best value of k using with-cluster-sum-of-squares and between-cluster-sum-of-squares.

   Ch is the ration of BSS/WSS.

Table 12.2 Data Set

METHOD NAME	DATA SET
1. Elbow Curve	Customer Segmentation, News purity filteration, Wireless sensor network.
2. Silhouette Curve	Classify the image based on color,Fuzzy clustering, Medical Images
3. Gap Statistics	Online phenotype discover, Blind source seperation, spike sorting
4. CH Index	Crude oil, Ionosphere and cancer dataset
5. DB Index	Cereals data, Web navigations

5. **DB-Index:** It will be constructed based on the ratio between within the cluster and between the cluster distance. It consider the smallest DB value as optimal value. In most of the data set it provides the acute value of K compare to any other algorithm.

Lets Compare these methods applicability in Table 12.2 [2].

Properties of k-Means

Properties of k-means algorithm: Clustering the labeled value or supervised clustering can be done by any labeled value which is common factor to differentiate the group such as gender for population, Product name in mall, rating of a cloths or disease name in patients are some of the example but clustering the unlabeled data set or unsupervised clustering is quiet different and complex while grouping due its nature of data such as networking traffic to find the malicious data flow, chromosomes behavior, image compression, Eruptions float and quality of soil and many more where there is no weight labeled to classify the data set.

k-means algorithm is a most popular algorithm which is more suitable for unsupervised clustering,as discussed in previous section the data are not having confined value which needs to be segmented based on multiple variables or a single variable. It comes under the categories of partitioning algorithm which resultant with linear time complexity which does not create any nested clustering hence overlapping between the cluster is overruled out in this clustering algorithm.

k-means algorithm can also conjugated with many other methods for varying performance, will discuss in detail in the next section.

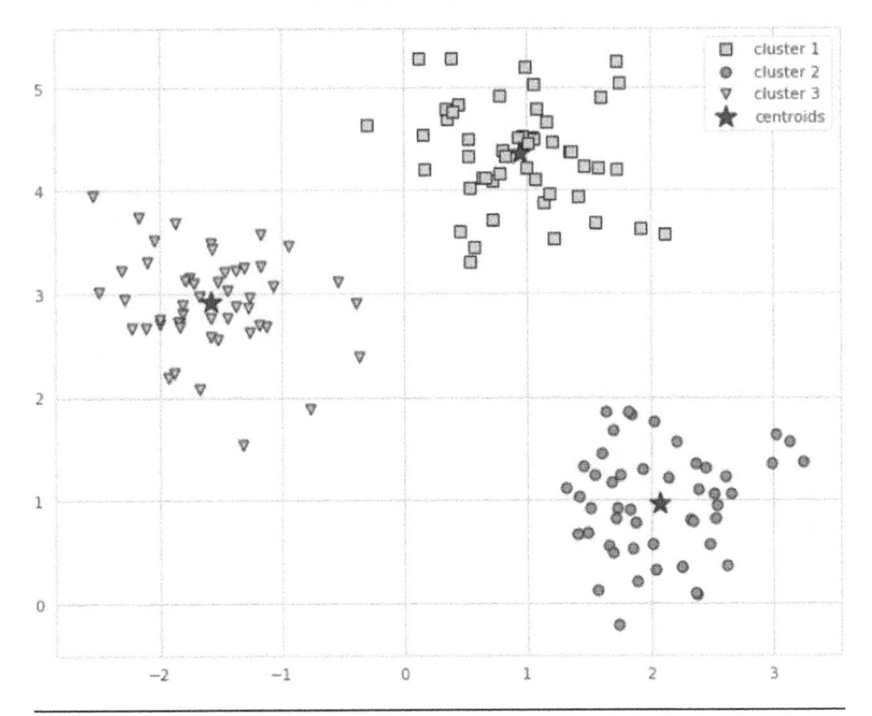

Figure 12.7 k-Means Clustering in Spherical Structure

Lets see the prominent features of k-means algorithm which provides the base line to choose it for our respective problem either for segmentation or compression.

Property 1: k-means *clustering is better in creating the* ***spherical structure*** *like cluster.* As shown in the image Figure 12.7, k-means clustering in a spherical structure gives more weight to bigger clusters than smaller ones and the smaller cluster may also be neglected.

Property 2: Vector quantization based single processing problem can opt the k-means due its basic characteristic which adopted Llody's algorithm especially for quantization. Consider a simple example where image compression need to carried out in which the similar pixel should be compressed by quantization, k-means is well suited for this as shown in the below image the color with similar density or pixel value can be easily grouped using k-means which give better output for heavy weighted ones. Since k-means is fast and have good trade off for the dataset which may not have large gap in the dataset. Figure 12.8 depicts the example of colored compression vector Quantization.

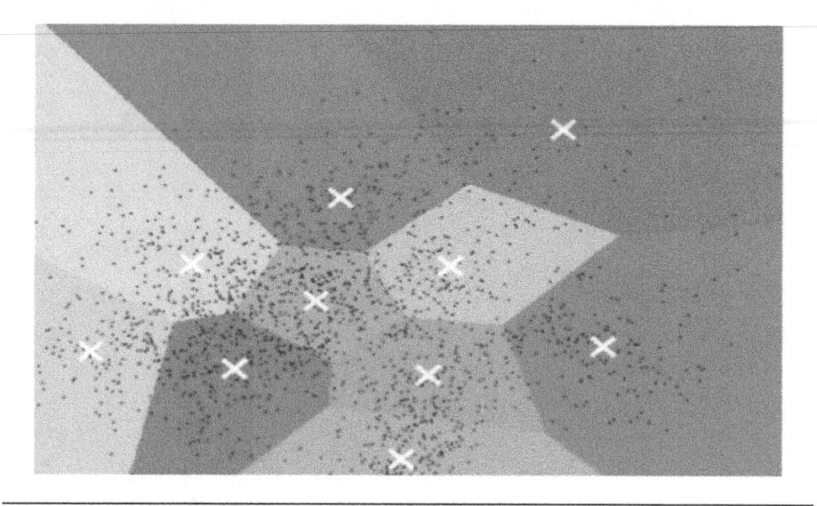

Figure 12.8 Vector Quantization [Each Data Point Will Be in the Voronoi Cell]

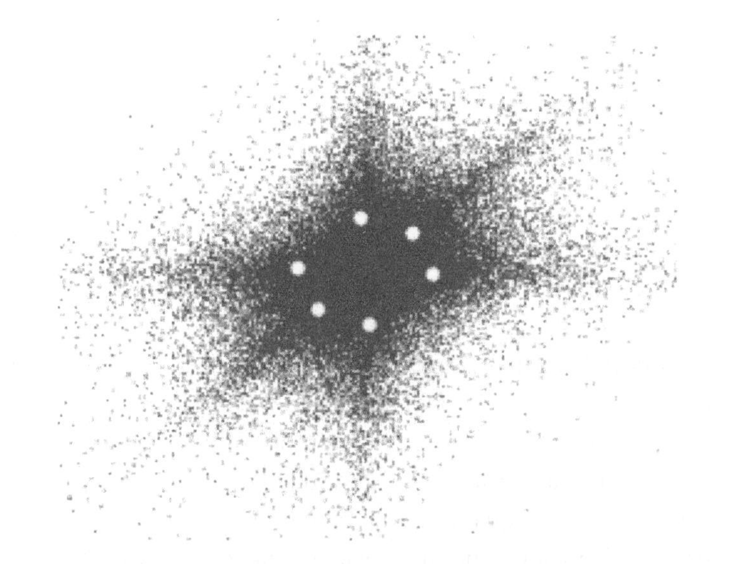

Figure 12.9 Mixing Matrix With k=6

Property 3: ***k-means in Linear independent component analysis,*** ICA is basically used in deep learning which is used to disclose the hidden values or random variable in data set. As discussed in [5], k-means played the major role in mixing matrix for filtering the data also the convergence rate is comparatively efficient. Consider the example discussed in [5] as depicted in Figure 12.9, k-means when integrated with ICA as come up as powerful tool to recover the

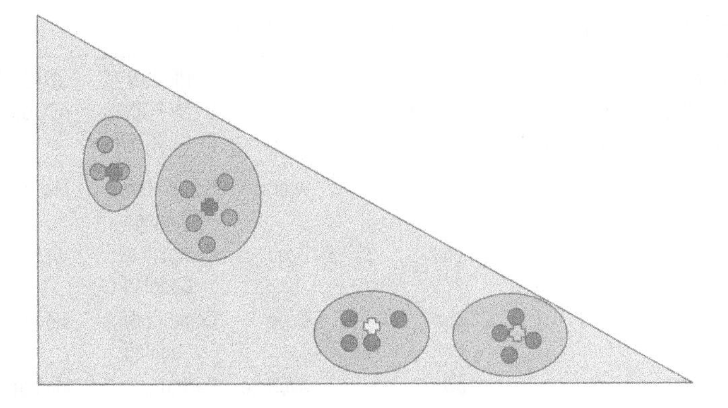

Figure 12.10 Clustered Network Notes Using k-Means

columns in mixing matrix and is also the best algorithm for dictionary learning.

Property 4: *Data points in a data set should be similar but it can be different from the each clusters.*

Consider an example of placing the router in the region based on the number of host and which are grouped based on either distance or rate of data transfer. Best part is to select the variable of distance than network traffic which are varying with time. k-means group the host which are with minimum distance between the host and plce the centroid to maintain the minimum distance between the cluster as shown below.

Consider the centroid c1,c2,c3,c4 where c1 is close to c4 and c2 is close to c3 then the connectivity looks as shown in the Figure 12.10.

As discussed above those 4 property plays a major role while implementing the clustering using k-means algorithm.

Compare the k-Means with Fuzzy C Mean, K Mediods, and k-Means++

In the previous section we learnt about k-means algorithm and its properties, but in some of the research work k-means is been compared with few other unsupervised algorithms which are overcome some of the drawback of k-means algorithms. Will study there concepts then compare with k-means.

a. ***Fuzzy C Mean:*** It is introduced by Bezdek[], this algorithm data points can be in more than one cluster based on the membership grade assigned to data points. This algorithm

Table 12.3 Comparison of Unsupervised Clustering Algorithms

ALGORITHMS/ PERFORMANCE PARAMETER	SQUARED ERROR MEASURED	QUICK CONVERGENCE	ROBUST	COHESION OF DATA POINTS IN CLUSTER	CLUSTERING ACCURACY
k-means	Yes	Average	Average	Completely Coupled	Medium
K-Mediod	No	Good	Good	Completely Coupled	High
k-means ++	Yes	Good	Good	Completely Coupled	Medium
Fuzzy C Mean	yes	Low	Average	Partially Coupled	Medium for large data set

combines with Gaussian and Expectation Maximization algorithm for more accuracy.

b. ***K Mediods:*** Mediods is like centroid k-means in which the datapoints should be minimum to this mediod to form cluster. As compared with k-mean, it aims to reduce the dissimilarity in data set by computing the pair wise distance perhaps time delay. If the value between mediod and data point is more than the swapping the data point to other mediod.
c. ***k-means ++:*** It overcomes the drawback of k-means algorithm which fails to initialize the centroid. In this algorithm the centroid chosen was initially in another cluster is the benefit in reducing the run-time for convergence.

With the above small description we would like to compare those 3 algorithm with K-means with the reference [22,23] (Table 12.3).

The above table helps to conclude the k-means algorithm is using squared distance error for measuring the accuracy and the data points are tightly coupled with few points to also be consider with respect to convergence but due its simplicity and optimal performance this can be used for large data set also.

k-Means Clustering in Software Defined Network (SDN)

Controlling the network components such as switches, nodes and also the gateways can be achieved using software by the new

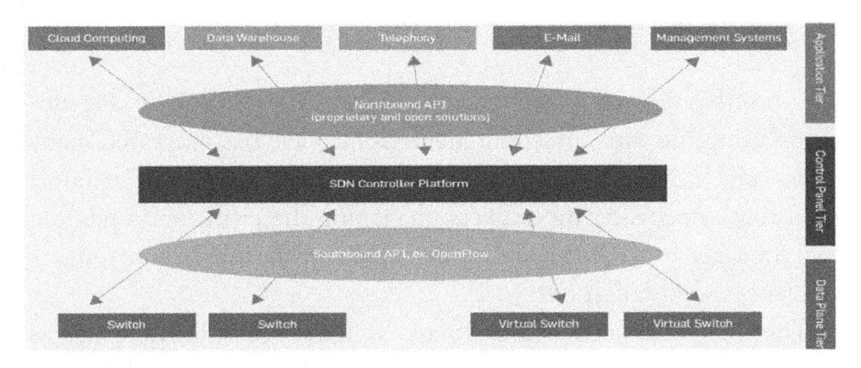

Figure 12.11 The Macrolevel SDN Architecture

technology called SDN. Internet of Thing and growing industries are making the world digitized and making the job simple with the power of internet from 2G to 5G.

A small introduction on SDN is as follows [24] (Figure 12.11),

- Northbound API's are logical application such as Firewalls, Routing algorithms, load balancer and so on which helps the network components to communicate with high level components.
- Southbound API's are used to communicate with switch fabrics, flows, protocols and network virtualization protocol.
- Controller is placed in intermediary between northbound and southbound application interface to manage, monitor and direct the network component centrally.

As we new southbound API's are the network components which will scale as the connectivity demand increases at the same time challenges with respect to scaling also place an major role in research which can be smoothed by clustering the hosts are switches and controllers.

Many research is heading towards to this challenges as discussed in [25–27] and many more, more interestingly the logical grouping of these nodes are also a part of solution in it.

k-means clustering algorithm is implemented by research to place the controller in optimal place and in [25], its is used to reduce the latency and to group the incoming network [26]. k-means is not only for placement but also used for detecting DDOS attack. Barki [27] as concluded though the k-means is less in accuracy but consume less time compared to other algorithms.

Exploring the k-Means Clustering in Resolving SDN Challenges

The proliferation of SDN such as centralizing the networking decision using the software component(Controller) has decreased many issues and improved the network performance in terms of resource utilization,processing the packets, managing the traffic and updating the firewalls to switches centrally, along with the benefits many challenges are branch out.

The challenges such as DDOS attack, Resource utilization such as bandwidth, traffic management and intercommunication delay between the controllers. Many researcher have proposed the novel algorithm and the mechanism to resolve and in most of the research clustering technique is adopted to resolve the issues among which k-means in been integrated by many. Lets see few of the studies to know the insight of those research.

Case 1- To reduce the latency in SDN: Placing the SDN controller in the appropriate place which helps to reduce the latency, as discussed in *Network Partitioning using k-means* [25] the latency compared to standard connectivity and clustering using k-means there is an decrease in latency for k=5,6 about achieved the latency around 2.3ms which is smaller than non clustering mechanism.

Case 2- DDoS attack monitoring in SDN: Felipe A. Lopes [28], has proven the k-means influence in monitoring DDoS attack and the implemented to detect the abnormal dataset using RYU application and successfully found monitor the network traffic for the attack.

Case 3- Traffic Monitoring: Classifying the network traffic based on the traffics and packet size. As discussed in [29],the k-means plays a major rule in classifying the traffic so as to place the controller by checking the saturation point for placing it. k-means also proved the best accuracy for traffic classification with the result of 90% accuracy reducing the squared error [29]. As experimented by Liu, the classification of network traffic for the data from 1000 users created the classification with the type of data transferred n which http protocol data

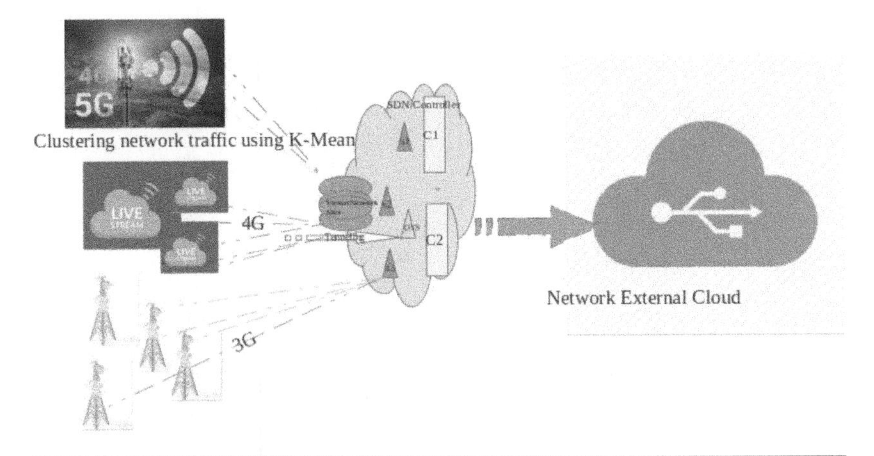

Figure 12.12 k-Means Clustering with SDN and 5G

transmission is high. Along with transformation steps log data is considered and applied k-means results with 90% of accuracy with the k=80.Installing the SDN to the same experiment is an novel approach for achieving more accuracy.

The role of k-means clustering is one of the strategical step to make the network more robust along with SDN and 5G network. As depicted below in Figure 12.12, k-means can play a major role in portioning the network.

As the new technology embed in current technology such as integration of 5G with SDN improves the Manageability, orchestration, services and resources. As Network Function Virtualization is helps to partition the network traffic by creating a virtual LAN and k-means can help in placing the controller in an efficient position as well classify the traffic to find the malicious attack or to find the type of traffic flowing toward network devices.

SDN Network Partitioning of Saren Topology-Based using k-Means

An small experiment conducted using mininet simulator to evidence the k-means role in placing the controller for managing the switches.

In this chapter, a small implementation is worked on Saren topologies dataset from Zootopology and computed the centroid for placing the controllers for better performance in terms of reducing delay is measured n millisecond.

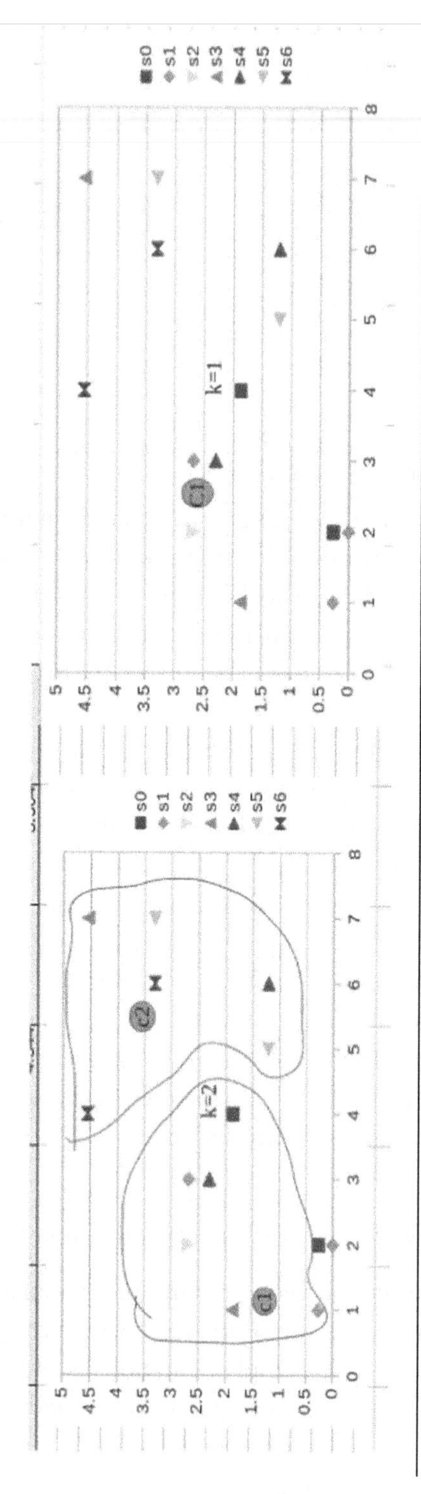

Figure 12.13 k-Means Clustering for Saren Topology

Methods	Latency
Tradition	15.22
K-mean k=1	12.85
k=2	8.43

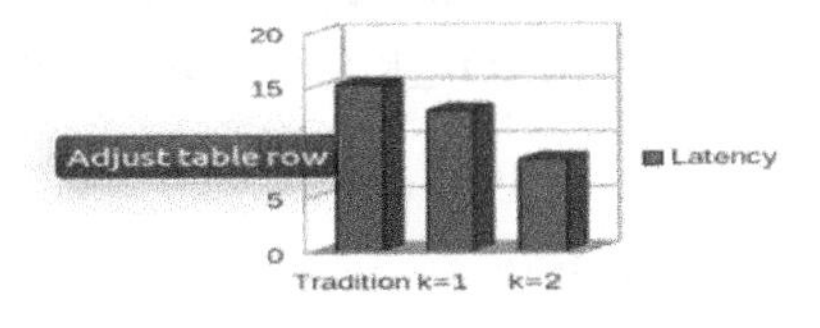

Figure 12.14 RTT Time after Applying k-Means Algorithm

As discussed in the k-means algorithm first step is assigning the dataset to an array as a delay vector for the 7 switches as mentioned in gml file. Since the switches as very less in number we computed for the value k=1,2 where the centroid value after applying euclidean distance formula is 2.6908 and 1.46, 3.5 respectively which is represented in the Figure 12.13.

We illustrated the effectiveness of k-means clustering and found that k=2 result the reduction of latency compares to k=1 in k-means and traditional approach. As the graph shows that when k=2 the delay is reduced around 15-20% and improve the response time (Figure 12.14).

Conclusion

The k-means algorithm is very popular and simple for implementation and is more suitable for the vector based data set. It is primarily using euclidean distance algorithm which is used to minimize the squared errors for more accuracy, through it started the journey in 1967 its powerful concept make it as ubiquity. The distance formula is varied for the suitable problem for better performance. Its efficiency and applicability is plethora but tried an small attempt with the example using SDN to highlight its applicability with best choice for partitioning the data points. Its also proven by many researcher about the efficiency of k-means and also we proved the improvement in network latency is reduced and increase in response time intern.

Code for implementing k-means is available in sklearn,

Sklearn---→Cluster----→k-means and refer the github given below for the source code of k-means in Python.

[https://github.com/scikit-learn/scikit-learn/blob/0fb307bf3/sklearn/cluster/_kmeans.py#L749]

References

[1] MacQueen, J. B. (1967). Some Methods for classification and Analysis of Multivariate Observations. Proceedings of 5th Berkeley Symposium on Mathematical Statistics and Probability.1. University of California Press. pp. 281–297. MR0214227.Zbl0214.46201. Retrieved 2009-04-07.

[2] Improved Clustering with Augmented k-means J. Andrew Howe, Independent Researcher, Riyadh, arXiv:1705.07592[stat.ML], May 2017.

[3] Krishna, K., and M. Murty. 1999. Genetic K-Means algorithm. IEEE Transactions on Systems, Man, and Cybernetics. Part B, Cybernetics: a Publication of the IEEE Systems, Man, and Cybernetics Society vol. 29 (3):pp. 433–439. doi: 10.1109/3477.764879.

[4] Li, Z., & Wang, H. (2019). Improving K-means method via shrinkage estimation and LVQ algorithm. Communications in Statistics - Simulation and Computation, vol. 50, pp. 1–16. doi:10.1080/03610918.2019.1620274

[5] Vinnikov, Alon; Shalev-Shwartz, Shai (2014). "K-means Recovers ICA Filters when Independent Components are Sparse"*(PDF). Proceedings of the International Conference on Machine Learning (ICML 2014).*

[6] "Supervised and Unsupervised Machine Learning Algorithms" by byJason BrownleeoninMachine Learning Algorithms

[7] Reffered "Mathworks" for implementation, https://in.mathworks.com/help/stats/cluster-analysis.html.

[8] MacQueen, J., 1967. Some Methods for Classification and Analysis of Multivariate Observations. In: Cam, L. M. L., Neyman, J. (Eds.), Proceedings of 5-th Berkeley Symposium on Mathematical Statistics and Probability. Vol. 1. University of California, Berkeley, Berkeley, USA, pp. 281–297.

[9] https://www.datanovia.com/en/lessons/determining-the-optimal-number-of-clusters-3-must-know-methods/

[10] Jenks, George F. 1967. "The Data Model Concept in Statistical Mapping", International Yearbook of Cartography, vol. 7: pp. 186–190.

[11] Vinnikov, Alon; Shalev-Shwartz, Shai (2014). "K-means Recovers ICA Filters when Independent Components are Sparse". *Proceedings of the International Conference on Machine Learning (ICML 2014).*

[12] Erman, J., Arlitt, M., Mahanti, A.: Traffic classification using clustering algorithms. In: Proceedings of the 2006 SIGCOMM

Workshop on Mining Network Data, MineNet 2006, pp. 281–286. ACM, New York (2006).

[13] Groenen, P. J. F., & Jajuga, K. (2001). Fuzzy clustering with squared Minkowski distances. Fuzzy Sets and Systems, vol. 120(2), pp. 227–237. doi:10.1016/s0165-0114(98)00403-5.

[14] Dibya Jyoti Bora, Dr. Anil Kumar Gupta. "Effect of Different Distance Measures on the Performance of K-Means Algorithm: An Experimental Study in MATLAB", (IJCSIT) International Journal of Computer Science and Information Technologies, Vol. 5 (2), 2014, pp. 2501–2506.

[15] M. K. Arzoo, A. Prof, and K. Rathod, "K-Means algorithm with different distance metrics in spatial data mining with uses of NetBeans IDE 8. 2," Int. Res. J. Eng. Technol., vol. 4, no. 4, pp. 2363–2368, 2017.

[16] "A K-Nearest Neighbor Technique for Brain Tumor Segmentation Using Minkowski Distance" , Xie, Xiaozhen, Journal of Medical Imaging and Health Informatics, Volume 8, Number 2, February 2018, pp. 180–185(6).

[17] Groenen, Patrick & Kaymak, Uzay & Rosmalen Joost. (2007). Fuzzy Clustering with Minkowski Distance Functions. 10.1002/9780470061190.ch3.

[18] Fuentes, Claudio, and George Casella. "Testing for the existence of clusters." SORT (Barcelona) vol. 33, 2 (2009): 115–157.

[19] In the article "Silhouette Analysis vs Elbow Method vs Davies-Bouldin Index: Selecting the optimal number of clusters for KMeans clustering", GEORGIOS DRAKOS, 4th march 2020. https://gdcoder.com/silhouette-analysis-vs-elbow-method-vs-davies-bouldin-index-selecting-the-optimal-number-of-clusters-for-kmeans-clustering/

[20] Maulik, U., & Bandyopadhyay, S. (2002). Performance evaluation of some clustering algorithms and validity indices. IEEE Transactions on Pattern Analysis and Machine Intelligence, 24(12), 1650–1654.doi:10.1109/tpami.2002.1114856

[21] Singh, A. K., Mittal, S., Malhotra, P., & Srivastava, Y. V. (2020). Clustering Evaluation by Davies-Bouldin Index(DBI) in Cereal data using K-Means. 2020 Fourth International Conference on Computing Methodologies and Communication (ICCMC). doi: 10.1109/iccmc48092.2020.iccmc-00057.

[22] B. Simhachalam, G. Ganesan, "Performance comparison of fuzzy and non-fuzzy classification methods", Egyptian Informatics Journal Volume 17, Issue 2, July 2016, Pages 183–188, ScienceDirect.

[23] Kalpit G. Soni,and Dr. Atul Patel, "Comparative Analysis of K-means and K-medoids Algorithm on IRIS Data", International Journal of Computational Intelligence Research ISSN 0973-1873 Volume 13, Number 5 (2017), pp. 899–906.

[24] Pushpa. J. Research and P. Raj, "Topology-based analysis of

performance evaluation of centralized vs. distributed SDN controller," 2018 IEEE International Conference on Current Trends in Advanced Computing (ICCTAC), Bangalore, 2018, pp. 1–8, doi: 10.1109/ICCTAC.2018.8370394.

[25] G. Wang, Y. Zhao, J. Huang, Q. Duan, and J. Li, "A k-means-based network partition algorithm for controller placement in software defined network," in 2016 IEEE Int. Conf. on Communications (ICC), pp. 1–6, May 2016

[26] Z. Lv et al., "An Optimizing and Differentially Private Clustering Algorithm for Mixed Data in SDN-Based Smart Grid," in IEEE Access, vol. 7, pp. 45773–45782, 2019, doi: 10.1109/ACCESS.2019.2909048.

[27] Barki, L., Shidling, A., Meti, N., Narayan, D. G., & Mulla, M. M. (2016). Detection of distributed denial of service attacks in software defined networks. 2016 International Conference on Advances in Computing, Communications and Informatics (ICACCI). doi:10.1109/icacci.2016.7732445.

[28] https://github.com/felipealencar/sdn-ddos-monitor

[29] Liu, Y., Li, W., & Li, Y.-C. (2007). Network Traffic Classification Using K-means Clustering. Second International Multi-Symposiums on Computer and Computational Sciences (IMSCCS 2007). doi:10.1109/imsccs.2007.52.

13

An Intelligent Deep Learning-Based Wireless Underground Sensor System for IoT-Based Agricultural Application

PRISCILLA RAJADURAI[1]
AND G. JASPHER W. KATHRINE[2]

[1]*St. Joseph's Institute of Technology, Chennai, India*
[2]*Karunya Institute of Technology and Sciences, Coimbatore, India*

Contents

DOI: 10.1201/9781003119838-13

Introduction

System Overview

Wireless underground sensor networks (WUSNs), which consist of wireless sensors buried underground, are a natural extension of the WSN phenomenon and have been considered as a potential field that will enable a wide variety of novel applications that were not possible before. Compared to the current underground sensor networks, which use wired communication methods for data retrieval, WUSNs have several remarkable merits, such as concealment, ease of deployment, timeliness of data, reliability, and coverage density. The realization of wireless underground communication and networking techniques will lead to potential applications in the fields of agriculture, border patrol, assisted navigation, sports field maintenance, intruder detection, and infrastructure monitoring. Many of these novel applications will be possible because WUSNs can provide localized and real-time data about a specific soil region and its surrounding area. Despite its potential advantages, several open research problems exist to make WUSNs feasible.

The main challenge is the realization of efficient and reliable underground wireless communication between buried sensors. Aspects such as temperature, weather, soil composition, soil moisture, and soil homogeneity directly impact the connectivity and communication success. In fact, underground communication is one of the few fields where the environment has a significant and direct impact on the communication performance. Hence, characterization of the wireless underground channel is essential for the proliferation of communication protocols for WUSNs. In this chapter, we introduce theoretical models to characterize the underground wireless channel. The models developed in this chapter characterize not only the propagation of EM wave in soil, but also other effects on the communication related to multipath effects, soil composition, water content, and burial depth. The results obtained from this formalization reveal that underground communication is severely affected by frequency and soil properties, and more specifically by the volumetric water content (VWC) of soil.

In addition to nodes buried underground, the existence of aboveground nodes is also necessary for WUSN due to data retrieval purposes. Accordingly, two communication types coexist in WUSNs. Underground-to-underground communication refers to the information exchange between buried sensors for network management and data relay purposes. The data in the network are then collected through underground-to-aboveground communication. This type of communication also includes command and control information from aboveground stations to underground. The classification also highlights significant differences on the challenges related to each option. Underground-to-aboveground communication presents better quality than underground-to-underground, since certain portion of the communication takes place over the air. On the other hand, underground-to-underground communication usually presents more difficult challenges for the design and the deployment of WUSNs. In this chapter, we focus on underground-to-underground communication for WUSNs.

Scope of the Project

The scope of this work is to show that the data can be transferred through soil wirelessly, to help the farmers to measure the moisture of the soil wirelessly and intimate the farmers through monitor, and to help maintain the soil moisture in the agriculture field. It also helps to water the soil when automatically the moisture of the soil reduces to certain level. It is done using WUSN sensor and the embedded C is needed in Arduino.

Literature Survey

Wireless underground sensor networks constitute one of the promising application areas of the recently developed WSN techniques. WUSN devices are deployed completely below the ground and do not require any wired connections. Each device contains all necessary sensors, memory, a processor, a radio, an antenna, and a power source. This makes their deployment much simpler than existing underground sensing solutions. Wireless communication within a dense substance such as soil or rock is, however, significantly more challenging than through air.

Underground Communications Networks

In agriculture, WSNs have been widely promoted as a means to improve on-farm yield and profitability through the provision of real-time or on-demand sensed data. Recent developments in communication techniques for WSN have resulted in a vast number of applications in agriculture. WSN is used mainly for management decisions of irrigation water resources, storage management of agricultural products, time determination of the crop harvest, characteristics of crop growth, forecast of fertilizer demand [1], and so on.

The WSN was a kind of network whose energy was limited, reasonable clusters were determined, and the energy model used in communication was given according to the number of nodes and characteristics of regional distribution [2]. Collection node system of farmland information based on WSN was designed [3].

The main difference between WUSNs and the terrestrial WSNs is the communication medium. The propagation characteristics of electromagnetic (EM) waves in soil and the significant differences between propagation in air prevent a straightforward characterization of the underground wireless channel [4]. The effects of variations in soil moisture are investigated through field measurement results. The theoretical analysis and the simulation results prove the feasibility of wireless communication in underground environment and highlight several important aspects in this field [5].

Dohare et al. [6] presented a comprehensive review of existing approaches for mine monitoring has. For the base protocol, most works have relied on Zigbee for its ease of deployment, low data rate (250 kb/s), substantial range and most importantly low power consumption when compared to other technologies, such as Wi-Fi, Bluetooth, and ultra-wideband communication.

Yarkan et al. [7] proposed that after a recent series of unfortunate underground mining disasters, the vital importance of communications for underground mining is underlined one more time. Establishing reliable communication is a very difficult task for underground mining due to the extreme environmental conditions. Until now, no single communication system exists, which can solve all of the problems and difficulties encountered in underground mine communications.

Donoghue [8] reviewed occupational health hazards in mining and found occupational exposure limits (OELs) for hazardous substances. The physical, compound, natural, ergonomic, and psychosocial word-related wellbeing perils of mining and related metallurgical forms are depicted. Mining remains a critical modern segment in many parts of the world and albeit significant advance has been made in the control of word-related wellbeing dangers; there remains space for further hazard decrease.

Kiran Lakkaraju et al. described that the number of attacks against large computer systems is currently growing at a rapid pace. Despite the best efforts of security analysts, large organizations are having trouble keeping on top of the current state of their networks. A tool called NVisionIP is designed to increase the security analyst's situational awareness. As humans are inherently visual beings, NVisionIP uses a graphical representation of a class-B network to allow analysts to quickly visualize the current state of their network.

WUSN in Crop Production

Agnelo et al. [9] studied the influence of the communication performance between the terrestrial nodes and the underground nodes in some factors, including antenna bandwidth of WSN nodes in 433 MHz frequency, the buried depth of nodes in the soil and water content of the soil. The temperature change in different position of feed warehouse was monitored through WSNs which were introduced by Green et al. [10]. Among all these technologies, the agriculture domain is mostly explored concerning the application of WSNs in improving the traditional methods of farming. Wireless sensor units (WSUs): Each WSU, deployed on-field, has four different type of components: application-specific sensors, processing unit, radio transceiver, and battery power.

X. Q. Yu et al. [11] conducted experimental measurements with commodity sensor motes at the frequency of 2.4 GHz and 433 MHz, respectively. Experiments are run to examine the received signal strength of correctly received packets and the packet error rate for a communication link. The tests show the potential feasibility of the WUSN with the use of powerful RF transceivers at 433-MHz frequency. Moreover, a classification for WUSN communication is shown. Finally, the effects of burial depth, internode distance, and volumetric water content of the soil on the signal strength and packet error rate in communication of WUSN are concluded.

Test Bed Development for WUSN

Xin Dong et al. [12] estimated the distortion of distributed soil moisture measurement using WUSNs. The main focus of the work is to analyze the impact of the environment and network parameters on the estimation distortion of the soil moisture. More specifically, the effects of rainfall, soil porosity, and vegetation root zone are investigated by exploiting a rainfall model, in addition to the effects of sampling rate, network topology, and measurement signal noise ratio. Spatio-temporal correlation is characterized to develop a measurement distortion model with respect to these factors. The evaluations reveal that with porous soil and shallow vegetation roots, high sampling rate is required for sufficient accuracy. In addition, the impact of rainfall on the estimation distortion has also been investigated. In a storm, which carries on a large area and lasts for a long time, the estimation distortion is decreased because of the increase in spatial correlation. Moreover, only few closest sensors are needed to estimate the values of an interested location. These findings are utilized to guide the design of WUSNs for soil moisture measurement to reduce the density of network and the sampling rate of the sensors, but at the same time maintain the performance of the system.

Underground is more complicated than the terrestrial environment, which contains not only air but also sand, rocks, and water with electrolyte. It is challenging to realize wireless communication in such complex environments. Underground communication solutions, while mostly shielded from the environment, require the ability to communicate through soil and adjust its parameters to adapt to dynamic changes in soil with the use of temperature, soil moisture, electric conductivity, and water potential sensors [13].

Soil Moisture Measurement

Ritsema Kuipers and Kleiboer [14–23] developed a new stand-alone wireless embedded network system recently for continuous monitoring of soil water contents at multiple depths. This information on the technical aspects of the system, including the applied sensor technology, the wireless communication protocols, the gateway station for data collection, and data transfer to an end user Web page for disseminating results, is provided to targeted audiences. Results from the first test of the network system are presented and discussed, including lessons learned so far and actions to be undertaken in the near future to improve and enhance the operability of this innovative measurement approach.

Underground Mining

Underground Mine Communications Serhan Yarkan et al. proposed that after a recent series of unfortunate underground mining disasters, the vital importance of communications for underground mining is underlined one more time. Establishing reliable communication is a very difficult task for underground mining due to the extreme environmental conditions. Until now, no single communication system exists which can solve all of the problems and difficulties encountered in underground mine communications. However, combining research with previous experiences might help existing systems improve, if not completely solve all of the problems. In this survey, underground mine communication is investigated. Major issues which underground mine communication systems must take into account are discussed. Communication types, methods, and their significance are presented.

Occupational Health Hazard in Mining M. Donoghue described the physical, chemical, biological, ergonomic, and psychosocial occupational health hazards of mining and associated metallurgical processes. Mining remains an important industrial sector in many parts of the world and although substantial progress has been made in the control of occupational health hazards, there remains room for further risk reduction. This applies particularly to traumatic injury hazards, ergonomic hazards, and noise. Vigilance is also required to ensure exposures to coal dust and crystalline silica remain effectively controlled.

Wireless Underground Sensor

Agriculture Field Monitoring Per Agnelo R. Silva et al., the WSN is now a day widely used to build decision support for overcoming many problems in real world. One of the most interesting fields having an increasing need of decision support systems is precision agriculture (PA). The purpose of this paper is to design and develop an agricultural monitoring system using WSN to increase the productivity and quality of farming without observing it for all the time manually. Temperature, humidity, and water levels are the most important factors for the productivity, growth, and quality of plants in agriculture. The temperature, humidity, and water level sensors are deployed to gather the temperature and humidity values. The sensor has to transmit the gathered information through the wireless communication network to the data server (cloud). The IoT gateway is in charge of the communication between the remote-controll serial devices and central control system. The farmers or the agriculture experts can observe the measurements from the web simultaneously. With the continuous monitoring of many environmental parameters, the grower can analyze the optimal environmental conditions to achieve maximum crop productiveness, for the better productivity, and to achieve remarkable energy savings.

Spatio-Temporal Soil Moisture Measurement Xin Dong et al. investigated the estimation distortion of distributed soil moisture measurement using WUSNs. The main focus of this paper is to analyze the impact of the environment and network parameters on the estimation distortion of the soil moisture. More specifically, the effects of rainfall, soil porosity, and vegetation root zone are investigated by exploiting a rainfall model, in addition to the effects of sampling rate, network topology, and measurement signal noise ratio. Spatio-temporal correlation is characterized to develop a measurement distortion model with respect to these factors. The evaluations reveal that with porous soil and shallow vegetation roots, high sampling rate is required for sufficient accuracy. In addition, the impact of rainfall on the estimation distortion has also been investigated. In a storm, which carries on a large area and lasts for a long time, the estimation distortion is decreased because of the increase in spatial correlation. Moreover, only few closest sensors are needed to estimate the values of an interested location. These findings are utilized to guide the design of WUSNs for soil moisture measurement to reduce the density of network and the sampling rate of the sensors but at the same time maintain the performance of the system.

Channel Model and Analysis for Wireless Underground Sensor Networks in Soil Medium Mehmet C. Vurana et al. WUSNs constitute one of the promising application areas of the recently developed WSN techniques. The main difference between WUSNs and the terrestrial WSNs is the communication medium. The propagation characteristics of electromagnetic (EM) waves in soil and the significant differences between propagation in air prevent a straightforward characterization of the underground wireless channel. To this end, in this paper, advanced channel models are derived to characterize the underground wireless channel and the foundational issues for efficient communication through soil are discussed. In particular, the underground communication channel is modeled considering not only the propagation of EM waves in soil, but also the other effects, such as multipath, soil composition, soil moisture, and burial depth. The propagation characteristics are investigated through simulation results of path loss between two underground sensors. Moreover, based on the proposed channel model, the resulting bit error rate is analyzed for different network and soil parameters.

System Analysis

Existing System

WSNs have vast number of applications. Among these applications, WUSNs are a high promising field. WUSNs consist of wireless sensors, which are buried underground. WUSNs will enable a wide variety of novel applications that were not possible with current wired underground monitoring techniques. Compared to the current underground sensor networks, which use wired communication methods for network deployment, WUSNs have several remarkable merits, such as concealment, ease of deployment, timeliness of data, reliability, and coverage density.

WUSNs can be used to monitor soil conditions maintaining parameters such as water content, mineral content, salinity, and temperature at optimal levels. Real-time knowledge of soil conditions is also useful for landscaping, where WUSNs can be combined with automatic sprinkler systems such that grass, trees, and flowers are watered only when needed.

Proposed System

WUSNs constitute one of the promising application areas of the recently developed WSN techniques. WUSN is a specialized kind of WSN that mainly focuses on the use of sensors at the subsurface region of the soil, that is, the top few meters of the soil. For a long time, this region has been used to bury sensors, usually targeting irrigation and environment monitoring applications, although without wireless communication capability; WUSNs promise to fill this gap and to provide the infrastructure for novel applications. The main difference between WUSNs and the terrestrial WSNs is the communication medium. In fact, the differences between the propagation of electromagnetic (EM) waves in soil and in air are so significant that a complete characterization of the underground wireless channel was only available recently.

In this project, we propose WUSN-based communication for agricultural land areas eliminating existing communication channels like wired and wireless sensor networks. The agricultural sensors like soil moisture and pH level of water of different parts of the agricultural land areas are obtained using WUSN circuit, which is emerged into the soil. Using soil as the medium of communication, the sensor values are transmitted and received in the base station suing WUSN transmitter and receiver. Once the base station receives the data, we uploading the agricultural land features into the cloud for backup and easy access by authorized persons.

System Requirements

A 64-bit PC can handle larger amounts of information than a 32-bit system. Since it can use a RAM of 4 GB, a 64-bit computer can be more responsive when running lots of programs at once. 1 GB of RAM is considered low and most desktop and laptop computers comes stock with at least 2 GB, but usually more. Most softwares define two sets of system requirements, i.e., minimum and recommended.

Software Requirements A software requirement specification is a description of a software system to be developed. It lays out functional and non-functional requirements and it also describes the operating system and tools used in the system and they are:

- MPLAB IDE
- Embedded C

Hardware Requirements Hardware specifications are technical description of the computer's components and capabilities, such as processor speed, model, manufacturer, etc. So the hardware components required for the proposed system are:

- PIC microcontroller
- Soil moisture sensor
- Humidity sensor
- LCD
- WUSN TX and RX

Module Description

- Interfacing sensors
- Programming microcontroller
- Data transmission through soil
- Visual basic
- Internet of things

Interfacing Sensor – Humidity Sensor

General Description A humidity sensor senses and measures both moisture and air temperature. The sensor is composed of two metal plates and contains a non-conductive polymer film between them. This film collects moisture from the air, which causes the voltage between the two plates to change. These voltage changes are converted into digital readings showing the level of moisture in the air.

Product Description

Humidity measurement can be done using electronic hygrometers. Electronic-type hygrometers or humidity sensors can be broadly divided into two categories, namely, capacitive sensing effects and resistive sensing effects. Resistive-type humidity sensors pick up changes in the resistance value of the sensor element in response to the change in the humidity. Thick film conductor of precious metals like gold and ruthenium oxide is printed and culminated in the shape of the comb to form an electrode. Then a polymeric film is applied on the electrode; the film acts as a humidity sensing film due to the existence of movable ions. Change in impedance occurs due to the change in the number of movable ions.

Features

- Input voltage: 5 v
- Output: analog(0–5 v)
- High performance
- Long-term stability
- Close tolerances
- Low cost

Applications

- Air conditioners
- Climate control for greenhouses
- Storage and warehouses
- Meteorological applications

Soil Moisture Sensor

General Description

Use the soil moisture sensor just as you would a traditional soil moisture meter with the additional advantages of automated data collection, graSoil moisturizing, and data analysis. Typical activities using our soil moisture sensor include acid-base titrations, studies of household acids and bases, monitoring of soil moisture change during chemical reactions or in an aquarium as a result of soil moisture to synthesis, investigations of acid rain and buffering, and analysis of water quality in streams and lakes.

Product Description

A soil moisture sensor, shown in Figure 13.1, is a device that measures the hydrogen-ion concentration (soil moisture) in a solution, indicating its acidity or alkalinity. In addition to measuring the soil moisture of liquids, it can also measure the moist and light level. The soil moisture sensor has an inbuilt meter to measure the light intensity. The soil moisture of a solution indicates how acidic or basic (alkaline) it is. The soil moisture term translates the values of the hydrogen ion concentration – which ordinarily ranges between about 1 and 10 × – 14 g equivalents per liter – into numbers between 0 and 14.

Features

- Supply voltage: 5 VDC
- Soil tester moisture – light – soil moisture
- Output: analog

Applications

- Acid-base titrations
- Analysis of water quality in streams and lakes.

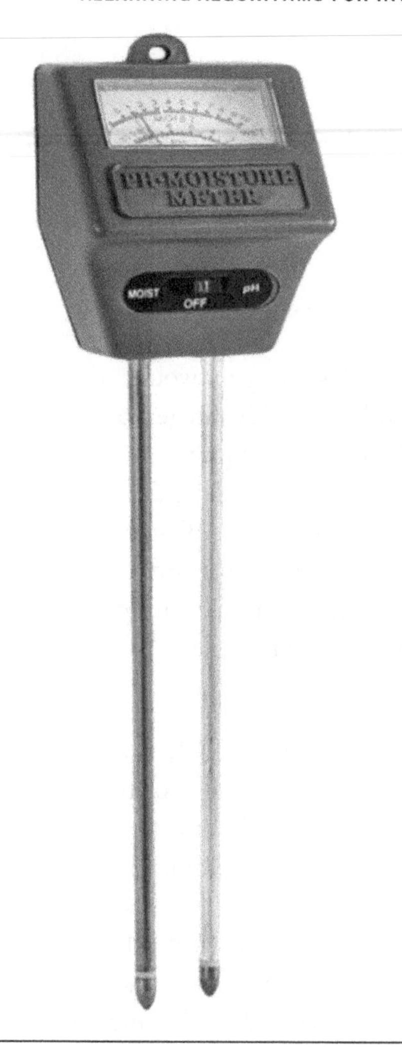

Figure 13.1 Moisture Detector Sensor

Programming Microcontroller

- Only 35 single-word instructions to learn.
- All single-cycle instructions except for program branches, which are two-cycles.
- Operating speed: DC – 20 MHz clock input DC – 200 ns instruction cycle.
- Up to 8K × 14 words of Flash program memory, Up to 368 × 8 bytes of data memory (RAM), Up to 256 × 8 bytes of EEPROM data memory is shown in Figure 13.2.

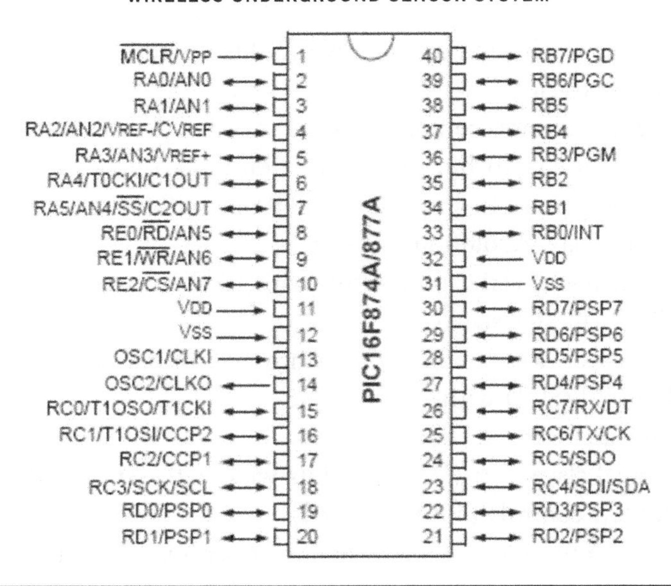

Figure 13.2 Pin Diagram of 877a

- Pin out compatible to other 28-pin or40/44-pin
- PIC16CXXX and PIC16FXXX microcontrollers are used in the microcontroller design as shown in Figure 13.3.

Perisoil Moistural Feature

- Timer0: 8-bit timer/counter with 8-bit prescaler
- Timer1: 16-bit timer/counter with prescaler, can be incremented during sleep via external crystal/clock

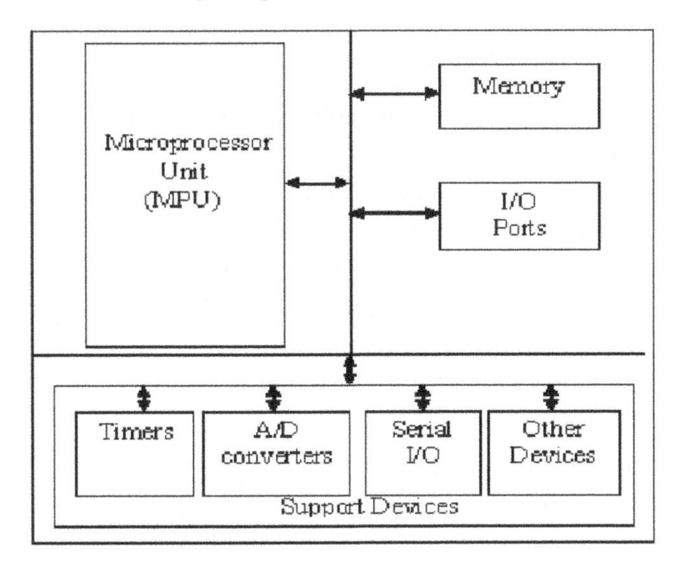

Figure 13.3 Microcontroller Design

- Timer2: 8-bit timer/counter with 8-bit period register, prescaler and postscaler
  - Two capture, compare, PWM modules
  - Capture is 16-bit, max. resolution is 12.5 ns
  - Compare is 16-bit, max. resolution is 200 ns
- PWM max. resolution is10-bit
- Synchronous serial port (SSP) with SPI™ (master mode) and I2C™ (master/slave)
- Universal synchronous asynchronous receiver transmitter (USART/SCI) with
- 9-bit address detection
- Parallel slave port (PSP) – 8 bits wide with external RD, WR and CS controls (40/44-pin only)
- Brown-out detection circuitry for brown-out reset (BOR)

Analog Features

- 10-bit, up to 8-channel analog-to-digital converter(A/D)
- Brown-out reset (BOR)
- Analog comparator module with:
  - Two analog comparators
  - Programmable on-chip voltage reference (VREF) module
  - Programmable input multiplexing from device inputs and internal voltage reference
  - Comparator outputs are externally accessible

Special Microcontroller Features

- 100,000 erase/write cycle enhanced Flash program memory typical
- 1,000,000 erase/write cycle data EEPROM memory typical
- Data EEPROM retention > 40 years
- Self-reprogrammable under software control
- In-Circuit Serial Programming™ (ICSP™) via two pins
- Single-supply 5-V ICSP
- Watchdog timer (WDT) with its own on-chip RC oscillator for reliable operation
- Programmable code protection
- Power-saving sleep mode

- Selectable oscillator options
- In-Circuit Debug (ICD) via two pins

CMOS Technology

- Low-power, high-speed Flash/EEPROM technology
- Fully static design
- Wide operating voltage range (2.0 V to 5.5 V)
- Commercial and industrial temperature ranges
- Low-power consumption

Pin Diagram

Device Overview This document contains device-specific information about the following devices:

- PIC16F873A
- PIC16F874A
- PIC16F876A
- PIC16F877A

PIC16F873A/876A devices are available only in 28-pin packages, while PIC16F874A/877A devices are available in 40-pin and 44-pin packages. All devices in the PIC16F87XA family share common architecture with the following differences:

- The PIC16F873A and PIC16F874A have one-half of the total on-chip memory of the PIC16F876A andPIC16F877A
- The 28-pin devices have three I/O ports, while the 40/44-pin devices have five
- The 28-pin devices have 14 interrupts, while the 40/44-pin devices have 15
- The 28-pin devices have five A/D input channels, while the40/44-pin devices have eight
- The parallel slave port is implemented only on the 40/44-pin devices

Block diagrams of the PIC16F874A/877A devices are provided in Figure 13.2. Additional information may be found in the PICmicro® Mid-Range Reference Manual (DS33023), which may be obtained from your local microchip sales representative or downloaded from the microchip website. The reference manual should be considered a complementary document to this data sheet and is highly recommended reading for a better understanding of the device architecture and operation of the Perisoil Moistural modules.

Memory Organization There are three memory blocks in each of the PIC16F87XA devices. The program memory and data memory have separate buses so that concurrent access can occur and is detailed in this section. The EEPROM data memory block is detailed in the Programming Microcontroller section and shown in Figure 13.4.

Data Memory Organization The data memory is partitioned into multiple banks which contain the general purpose registers and the special function registers (SFRs) as shown in Figure 13.4. Bits RP1 (Status<6>) and RP0 (Status<5>) are the bank select bits. Each bank extends up to 7 Fh (128 bytes). The lower locations of each bank are reserved for the SFRs. Above the SFRs are general purpose registers, implemented as static RAM. All implemented banks contain SFRs.

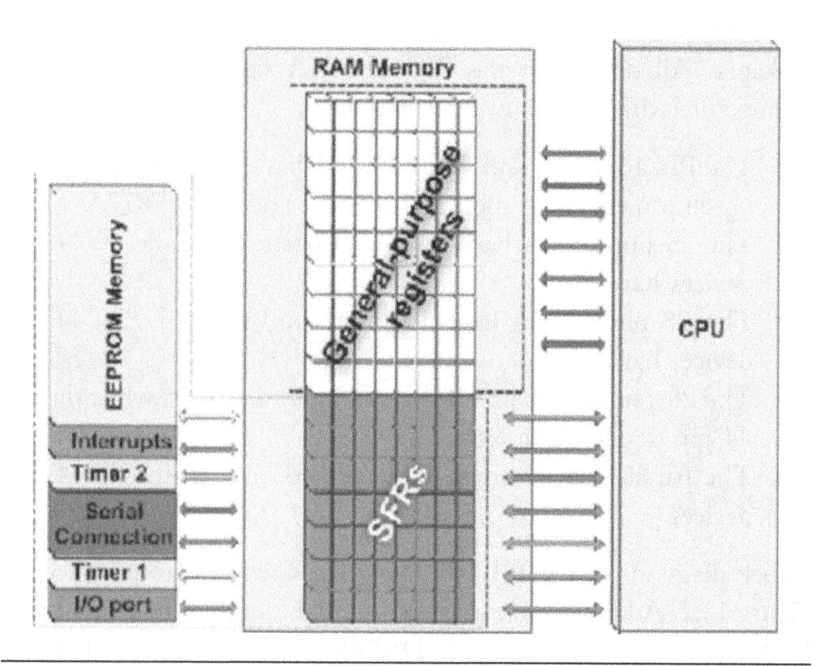

Figure 13.4 Memory Organization of a Controller

Some frequently used SFRs from one bank may be mirrored in another bank for code reduction and quicker access.

I/O ports: Some pins for these I/O ports are multiplexed with an alternate function for the Perisoil Moistural features on the device. In general, when a Perisoil Moistural is enabled, that pin may not be used as a general purpose I/O pin. Additional information on I/O ports may be found in the PICmicro™ Mid-Range Reference Manual (DS33023).

PORTA and the TRISA Registers PORTA is a 6-bit wide, bidirectional port. The corresponding data direction register is TRISA. Setting a TRISA bit (= 1) will make the corresponding PORTA pin an input (i.e., put the corresponding output driver in a high-impedance mode). Clearing a TRISA bit (= 0) will make the corresponding PORTA pin an output (i.e., put the contents of the output latch on the selected pin). Reading the PORTA register reads the status of the pins, whereas writing to it will write to the port latch. All write operations are read-modify-write operations. Therefore, a write to a port implies that the port pins are read; the value is modified and then written to the port data latch.

Pin RA4 is multiplexed with the Timer0 module clock input to become the RA4/T0CKI pin. The RA4/T0CKI pin is a Schmitt Trigger input and an open-drain output. All other PORTA pins have TTL input levels and full CMOS output drivers. Other PORTA pins are multiplexed with analog inputs and the analog VREF input for both the A/D converters and the comparators. The operation of each pin is selected by clearing/setting the appropriate control bits in the ADCON1 and/or CMCON registers. The TRISA register controls the direction of the port pins even when they are being used as analog inputs. The user must ensure that the bits in the TRISA register are maintained set when using them as analog inputs.

PORTB and the TRISB Registers PORTB is an 8-bit wide, bidirectional port. The corresponding data direction register is TRISB. Setting a TRISB bit (= 1) will make the corresponding PORTB pin an input (i.e., put the corresponding output driver in a high-impedance mode). Clearing a TRISB bit (= 0) will make the corresponding PORTB pin an output (i.e., put the contents of the output latch on the selected pin). Three pins of PORTB are multiplexed with the in-circuit debugger and low-voltage programming function: RB3/PGM, RB6/PGC and RB7/PGD. " Each of the PORTB pins has a weak internal pull-up. A single control bit can turn on all the pull-ups. This is performed by clearing bit RBPU (OPTION_REG<7>). The weak pull-up is automatically turned off when the port pin is configured as an output. The pull-ups are disabled on a power-on reset.

This interrupt can wake the device from sleep. The user, in the interrupt service routine, can clear the interrupt in the following manner:

a. Any read or write of PORTB. This will end the mismatch condition.
b. Clear flag bit RBIF.

A mismatch condition will continue to set flag bit RBIF. Reading PORTB will end the mismatch condition and allow flag bit RBIF to be cleared. The interrupt-on-change feature is recommended for wake-up on key depression operation and operations where PORTB is only used for the interrupt-on-change feature. Polling of PORTB is not recommended while using the interrupt-on-change feature. This interrupt-on-mismatch feature, together with software configurable pull-ups on these four pins, allow easy interface to a keypad and make it possible for wake-up on key depression.

PORTC and the TRISC Register: PORTC is an 8-bit wide, bi-directional port. The corresponding data direction register is TRISC. Setting a TRISC bit (= 1) will make the corresponding PORTC pin an input (i.e., put the corresponding output driver in a high-impedance mode). Clearing a TRISC bit (= 0) will make the corresponding PORTC pin an output (i.e., put the contents of the output latch on the selected pin). PORTC is multiplexed with several per Soil Moistureral functions. PORTC pins have Schmitt Trigger input buffers. When the I2C module is enabled, the PORTC<4:3> pins can be configured with normal I2C levels, or with SMBus levels, by using the CKE bit (SSPSTAT<6>). When enabling Perisoil Moistural functions, care should be taken in defining TRIS bits for each PORTC pin. Some Perisoil Moistural override the TRIS bit to make a pin an output, while other Perisoil Moistural override the TRIS bit to make a pin an input. Since the TRIS bit override is in effect, while the Perisoil Moistural is enabled, read-modify write instructions (BSF, BCF, and XORWF) with TRISC as the destination should be avoided. The user should refer to the corresponding Perisoil Moistural section for the correct TRIS bit settings.

PORTD and TRISD Registers PORTD is an 8-bit port with Schmitt Trigger input buffers. Each pin is individually configurable as an input or output. PORTD can be configured as an 8-bit wide microprocessor port (parallel slave port) by setting control bit, PSPMODE (TRISE<4>). In this mode, the input buffers are TTL.

PORTE and TRISE Registers PORTE has three pins (RE0/RD/AN5, RE1/WR/AN6, and RE2/CS/AN7) which are individually configurable as inputs or outputs. These pins have Schmitt Trigger input buffers. The PORTE pins become the I/O control inputs for the microprocessor port when bit PSPMODE (TRISE<4>) is set. In this mode, the user must make certain that the TRISE<2:0> bits are set and that the pins are configured as digital inputs. Also, ensure that ADCON1 is configured for digital I/O. In this mode, the input buffers are TTL.

The TRISE register which also controls the parallel slave port operation. PORTE pins are multiplexed with analog inputs. When selected for analog input, these pins will read as "0"s. TRISE controls the direction of the RE pins, even when they are being used as analog inputs. The user must make sure to keep the pins configured as inputs when using them as analog inputs.

Features of Microcontroller Microcontroller is the simplest computer processor which is used as the "brain" of the future system. Depending on the need, the manufacturer can extend the memory, converters, timers, input/output lines, etc. which are packaged in a standard format. Simple software can be used to control the microcontroller. Hence microcontroller has become man's invisible companion.

Their incredible simplicity and flexibility conquered us a long time ago and if you try to invent something about them, you should know that you are probably late, someone before you has either done it or at least has tried to do it.

The following things have had a crucial influence on development and success of the microcontrollers:

- Powerful and carefully chosen electronics embedded in the microcontrollers can independently or via input/output devices (switches, push buttons, sensors, LCD displays, relays, etc.), control various processes and devices, such as industrial automation, electric current, temperature, engine performance, etc.
- Very low prices enable them to be embedded in such devices in which, until recent time it was not worthwhile to embed anything. Thanks to that, the world is overwhelmed today with cheap automatic devices and various "smart" appliances.
- Prior knowledge is hardly needed for programming. It is sufficient to have a PC (software in use is not demanding at all and is easy to learn) and a simple device (called the programmer) used for "loading" ready-to-use programs into the microcontroller.
- Even though there is a large number of different types of microcontrollers and even more programs created for their use only, all of them have many things in common. Hence if one is handled, then all microcontroller devices can be handled.

Inside a Microcontroller

In a microcontroller, all the operations within the microcontroller are performed at high speed and quite simply, but the microcontroller itself would not be so useful if there are no special circuits which make it complete. In continuation, we are going to call your attention to them.

Read Only Memory (ROM)

- Read only memory (ROM) is a type of memory used to permanently save the program being executed. The size of the program that can be written depends on the size of this memory.

- ROM can be built in the microcontroller or added as an external chip, which depends on the type of the microcontroller. Both options have disadvantages.
- If ROM is added as an external chip, the microcontroller is cheaper and the program can be considerably longer. At the same time, a number of available pins are reduced as the microcontroller uses its own input/output ports for connection to the chip.
- The internal ROM is usually smaller and more expensive, but leaves more pins available for connecting to Perisoil Moistural environment. The size of ROM ranges from 512 B to 64 KB.

Random Access Memory (RAM)

- Random access memory (RAM) is a type of memory used for temporary storing data and intermediate results created and used during the operation of the microcontrollers.
- The content of this memory is cleared once the power supply is off. For example, if the program performs an addition, it is necessary to have a register standing for what in everyday life is called the "sum."
- For that purpose, one of the registers in RAM is called the "sum" and used for storing results of addition. The size of RAM goes up to a few KBs.

Electrically Erasable Programmable ROM (EEPROM)

- The EEPROM is a special type of memory not contained in all microcontrollers.
- Its contents may be changed during program execution (similar to RAM), but remains permanently saved even after the loss of power (similar to ROM).
- It is often used to store values, created and used during operation (such as calibration values, codes, values to count up to etc.), which must be saved after turning the power supply off.
- A disadvantage of this memory is that the process of programming is relatively slow. It is measured in milliseconds.

Memory Organization of Microcontroller

Special Function Register (SFR)

Special function registers are part of RAM memory. Their purpose is predefined by the manufacturer and cannot be changed therefore. Since their bits are soil moisture connected to particular circuits within the microcontroller, such as A/D converter, serial communication module etc., any change of their state directly affects the operation of the microcontroller or some of the circuits.

For example, writing zero or one to the SFR controlling, an input/output port causes the appropriate port pin to be configured as input or output. In other words, each bit of this register controls the function of one single pin.

Program Counter: Program Counter is an engine running the program and points to the memory address containing the next instruction to execute. After each instruction execution, the value of the counter is incremented by 1.

However, the value of the program counter can be changed at any moment, which causes a "jump" to a new memory location. This is how subroutines and branch instructions are executed.

After jumping, the counter resumes even and monotonous automatic counting +1, +1, +1 and so on.

Central Processing Unit (CPU) As its name suggests, this is a unit which monitors and controls all processes within the microcontroller and the user cannot affect its work.

It consists of several smaller subunits, of which the most important are:

> Instruction decoder is a part of the electronics which recognizes program instructions and runs other circuits on the basis of that. The abilities of this circuit are expressed in the "instruction set" which is different for each microcontroller family.

Arithmetical logical unit (ALU) performs all mathematical and logical operations upon data.

Accumulator is an SFR closely related to the operation of ALU. It is a kind of working desk used for storing all data upon which some operations should be executed (addition, shift, etc.).

It also stores the results ready for use in further processing. One of the SFRs, called the Status Register, is closely related to the accumulator, showing at any given time the "status" of a number stored in the accumulator (the number is greater or less than zero etc.). A bit is just a word invented to confuse novices at electronics. Joking aside, this word in practice indicates whether the voltage is present on a conductor or not. If it is present, the appropriate pin is set to logic one (1), i.e., the bit's value is 1.

Otherwise, if the voltage is 0, the appropriate pin is cleared (0), i.e. the bit's value is 0. It is more complicated in theory where a bit is referred to as a binary digit, but even in this case, its value can be either 0 or 1.

Data Transmission through Soil The concept of WUSN was introduced in wireless communication lab at the Georgia Institute of Technology in 2006. WUSNs have been investigated in many contexts recently, but research reports of WUSNs in agricultural application are few. I put forward the concept of WUSN for the first time and analyzed path loss of the underground wireless channel under the different soil composition. Besides, the key issues that should be considered were put forward in the WUSN communication architecture and WUSN design who studied the channel model of electromagnetic wave transmission in soil, analyzed path loss, bit error rate, maximum transmission distance, water content test error of electromagnetic wave multipath transmission, etc. under the main influence factors such as the soil composition, soil volumetric water content and node burial depth, node distance and sensor frequency studied the effect of antenna bandwidth of wireless sensor network node at 433 MHz frequency.

The buried depth of nodes in the soil 15 cm and 35 cm, and water content of the soil 9.5% and 37.3% on communication between aboveground node and underground node developed the near surface WUSN system used for golf course which included acquisition nodes, sink nodes, relay nodes, and a gateway node. The gateway node controlled data storage and transmission of the sink node could connect with computer or GPRS module through the RS232 interface. Like sink node, gateway node could be remotely controlled through DDI researched wireless signal attenuation of Zigbee wireless transceiver module by using soil column.

In the wireless underground sensor network AG-UG and UG-AG communication, sink node is deployed on the ground surface that is perpendicular to underground node, underground node burial depth changes every 10 cm within 10 cm to 100 cm, soil water content in the range of 5% to 30%, a total of six levels, the influence characteristics of node burial depth, and soil water content on received signal strength are measured in AG-UG and UG-AG communication. AG-UG and UG-AG communication test model.

In the actual WUSN communication, is not only communication that exists between the sink node and underground node, but also communication between underground nodes is essential, and is particularly important. In UG-UG communication, transmitting node and the receiving node burial depth are fixed 40 cm, soil water content changes from 5% to 30%, separated in six levels, the horizontal inter-nodes distance changes from 100 cm to 1000 cm, at 10 levels, and the influence characteristics of horizontal inter-nodes distance and soil water content on received signal strength are measured in UG-UG communication.

Visual Basic Visual Basic is engineered for productively building type-safe and object-oriented applications. Visual Basic enables developers to target Windows, web, and mobile devices. As with all languages targeting the Microsoft.NET Framework, programs written in Visual Basic benefit from security and language interoperability.

Visual Basic (VB) is a programming environment from Microsoft in which a programmer uses a graSoil Moistureical user interface (GUI) to choose and modify preselected sections of code. VB also includes advanced features – concepts and structures which allow programs to be adapted for use with the Internet. Visual Basic will not run on operating systems other than Windows and on machines with non-Intel compatible processors.

Internet of Things (IoT) The Internet of things (IoT) is the internetworking of physical devices, vehicle, buildings, and other items—embedded with electronics, software, sensors, actuators, and network connectivity that enable these objects to collect and exchange data. In 2013, the Global Standards Initiative on Internet of Things (IoT-GSI) defined the IoT as "the infrastructure of the information society." The IoT allows objects to be sensed or controlled remotely across existing network infrastructure, creating opportunities for more direct integration of the physical world into computer-based systems, and resulting in improved efficiency, accuracy, and economic benefit in addition to reduced human intervention. When IoT is augmented with sensors and actuators, the technology becomes an instance of the more general class of cyber-physical systems, which also encompasses technologies such as smart grids, smart homes, intelligent transportation and smart cities. Each thing is uniquely identifiable through its embedded computing system but is able to interoperate within the existing Internet infrastructure. Typically, IoT is expected to offer advanced connectivity of devices, systems, and services that goes beyond machine-to- machine (M2M) communications and covers a variety of protocols, domains, and applications. The interconnection of these embedded devices (including smart objects) is expected to user in automation in nearly all fields, while also enabling advanced applications like a smart grid, and expanding to areas such as smart cities. "Things," in the IoT sense, can refer to a wide variety of devices, such as heart monitoring implants, biochip transponders on farm animals, electric clams in coastal waters, automobiles with built-in sensors, DNA analysis devices for environmental/food/pathogen monitoring, or field operation devices that assist firefighters in search and rescue operations. Legal scholars suggest to look at "Things" as an "inextricable mixture of hardware, software, data, and service." These devices collect useful data. Current market examples include home automation (also known as smart home devices) such as the control and automation of lighting, heating (like smart thermostat), ventilation, air conditioning (HVAC) systems, and appliances such as washer/dryers, robotic vacuums, air purifiers, ovens or refrigerators/freezers that use Wi-Fi for remote monitoring.

System Architecture

Architecture Description

We developed the magnetic induction (MI) waveguide technique for the wireless communications in WUSNs. The MI waveguide consists of a series of relay coils between two underground transceivers. The communications are achieved by magnetic induction between two adjacent coils. These relay coils do not consume extra energy and the cost is neglectable. The MI waveguide can solve the two major problems of the EM technique in soil medium:

1. By using MI waveguides, the feasible communication range between two transceivers can achieve nearly 100 meters.
2. The MI channel conditions remain constant, since the soil medium causes little variation in the attenuation rate of magnetic fields.

Wireless Underground Sensor Design Architecture

As shown in Figure 13.5, the data from any soil can be collected through WUSN and the analysis of the soil can be used to predict the crops and the time for watering. A sprinkler system can also be combined in this proposed scheme to automate watering of crops based on the water content of the soil.

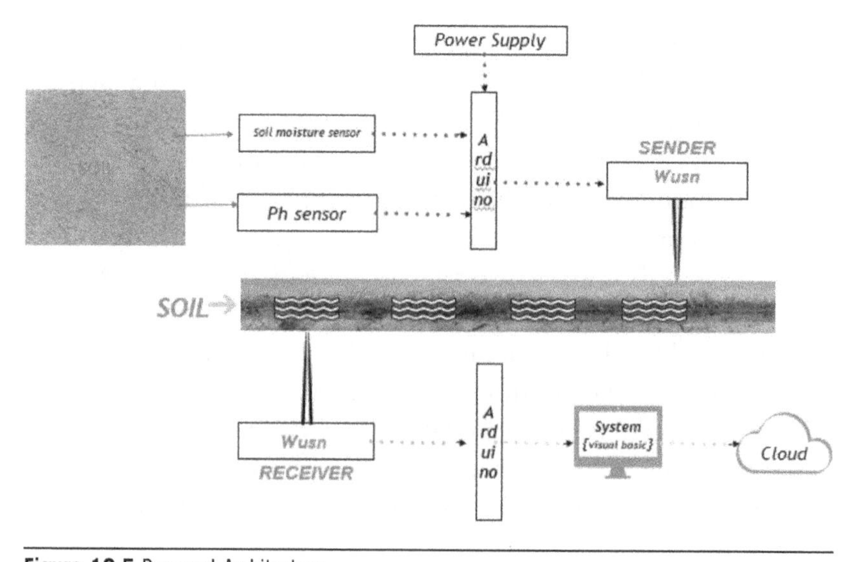

Figure 13.5 Proposed Architecture

Components of WUSN

Cloud Cloud computing is the on-demand availability of computer system resources, especially data storage (cloud storage) and computing power, without direct active management by the user. The term is generally used to describe data centers available to many users over the Internet.

Arduino Arduino is an open-source platform used for building electronics projects as shown in Figure 13.6. Arduino consists of both a physical programmable circuit board (often referred to as a microcontroller) and a piece of software, or IDE (integrated development environment) that runs on your computer, used to write and upload computer code to the physical board.

The Arduino platform has become quite popular with people just starting out with electronics, and for good reason. Unlike most previous programmable circuit boards, the Arduino does not need a separate piece of hardware (called a programmer) in order to load new code onto the board – you can simply use a USB cable. Additionally, the Arduino IDE uses a simplified version of C++, making it easier to learn to program. Finally, Arduino provides a standard form factor that breaks out the functions of the microcontroller into a more accessible package.

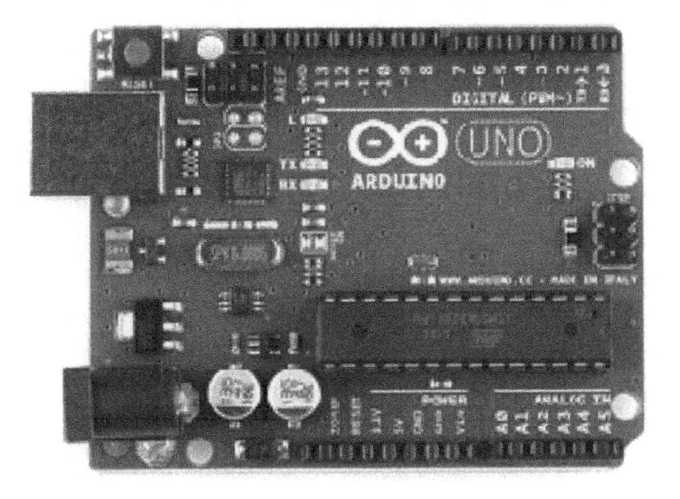

Figure 13.6 Arduino

pH Sensor A pH sensor is one of the most essential tools that's typically used for water measurements. This type of sensor is able to measure the amount of alkalinity and acidity in water and other solutions. pH sensors and transmitters are used in many industries such as chemicals, water and wastewater, food and beverage, pharmaceuticals, power plants, primaries and oil and gas. The selection of the sensor depends on the application. Choose the glass sensor's diaphragm and reference system according to your needs.

Soil Moisture Sensor A soil moisture sensor measures the quantity of water contained in a material, such as soil on a volumetric or gravimetric basis. To obtain an accurate measurement, a soil temperature sensor is also required for calibration. The soil moisture sensor uses capacitance to measure the water content of soil by measuring the dielectric permittivity of the soil, which is a function of the water content. Simply insert this rugged sensor into the soil to be tested, and the volumetric water content of the soil.

Transmitter and Receiver The WUSN transmitter is an electronic device which produces radio waves with an antenna. Transmitter itself generates a radio frequency alternating current, which is applied to the antenna. The transmitter is buried in the soil and gets energized when the data are passed. The receiver is the destination of the message. The receiver's task is to interpret the sender's message, both verbal and nonverbal, with as little distortion as possible. The process of interpreting the message is known as decoding.

Classification Based on Soil As shown in Table 13.1, the moisture content of the soil and its retention capabilities can confirm the type of soil. This needs no further analysis and the crops can be decided on the basis of soil type.

Table 13.2 gives the names of crops that can be grown in farms based on the soil type.

Table 13.1 Soil Type

TYPES OF SOIL	HIGH WATER RANGE	NORMAL	BELOW THE LEVEL
Fine (clay)	80–100	60–80	below 60
Medium (loamy)	88–100	70–88	below 70
Coarse (sandy)	90–100	80–90	below 80

Table 13.2 Crops for Soil Types

SOIL TYPE	CROP
Clay soil	Beans, cauliflower, cabbage
Loamy soil	Oil seed, sugarcane, wheat
Sandy soil	Maize, millets, wheat

Table 13.3 Result Obtained

DAYS	WATER LEVEL
1	49
2	48
3	49
4	40
5	37
6	35
7	30
8	20
9	15
10	39

Result The proposed architecture was implemented and the soil moisture was continuously monitored for 10 days. Based on the results obtained, as shown in Table 13.3, the analysis was done.

Based on the results obtained in Table 13.3, it is confirmed that the soil under consideration is clay type and the crops as shown in Table 13.2 for clay soil can be grown to get a better yield. The time taken for transmitting and receiving of the information from the sensors is a few seconds, as shown in Figure 13.7.

Conclusion

To conclude, the successful deployment of an IoT-enabled large-scale urban-area WUSN has been implemented. It covers the wide area, which is a typical urban environment with various buildings and human activities. All the sensor nodes are buried underground without interfering with the aboveground activities. Despite the extreme environment, it is deployed in underground, urban environment, and extremely cold winters.

LogID	DATA	Logdate	LogTime
1	Moisture:_PH:	07/13/2020	23:06:58
2	Moisture:_PH:	07/13/2020	23:07:02
3	Moisture:499_PH:0	07/13/2020	23:07:12
4	Moisture:499_PH:0	07/13/2020	23:07:22
5	Moisture:498_PH:0	07/13/2020	23:07:32
6	Moisture:498_PH:0	07/13/2020	23:07:42
7	Moisture:498_PH:0	07/13/2020	23:07:52
8	Moisture:496_PH:0	07/13/2020	23:08:02
9	Moisture:496_PH:0	07/13/2020	23:08:12
10	Moisture:496_PH:0	07/13/2020	23:08:22
11	Moisture:496_PH:102	07/13/2020	23:08:32
12	Moisture:496_PH:97	07/13/2020	23:08:42
13	Moisture:496_PH:111	07/13/2020	23:08:52
14	Moisture:496_PH:122	07/13/2020	23:09:02
15	Moisture:496_PH:131	07/13/2020	23:09:12
16	Moisture:496_PH:140	07/13/2020	23:09:22
17	Moisture:496_PH:145	07/13/2020	23:09:32
18	Moisture:496_PH:151	07/13/2020	23:09:42
19	Moisture:496_PH:155	07/13/2020	23:09:52
20	Moisture:496_PH:160	07/13/2020	23:10:02
21	Moisture:496_PH:163	07/13/2020	23:10:12
22	Moisture:496_PH:166	07/13/2020	23:10:22
23	Moisture:496_PH:167	07/13/2020	23:10:32
24	Moisture:496_PH:0	07/13/2020	23:10:42
25	Moisture:496_PH:0	07/13/2020	23:10:52

Figure 13.7 Time Analysis

WUSN has proven its value in real-time soil property monitoring. More importantly, we also implemented a cloud-based open platform, so the sensing results are open to public access. In this way, proposed system provides long-term continuous monitoring of the soil properties.

References

[1] E. Berman, G. Calinescu, C. Shah, and A. Zelikovsky., "Power efficient monitoring management in sensor networks," in *Proceedings of IEEE Wireless Communication and Networking Conference (WCNC04)*, Atlanta. USA, 2004.

[2] L. Li and X. M. Wen, "Energy efficient optimization of clustering algorithm in wireless sensor network," *J. Electron. Inform. Technol.*, vol. 30, no. 4, pp. 966–969, 2008.

[3] Y. H. Cai, G. Liu, L. Li, and H. Liu, "Design and test of nodes for farmland data acquisition based on wireless sensor network," *Chinese Soc. Agric. Eng.*, vol. 25, no. 4, pp. 176–178, 2009.

[4] M. C. Vuran and Ian F. Akyildiz, "Channel model and analysis for wireless underground sensor networks in soil medium,"in *Physical Communication*, vol. 3, no. 4, pp. 245–254, December 2010.

[5] M. C. Vuran and I. F. Akyildiz, "Channel model and analysis for wireless underground sensor networks in soil medium," *Physical Communication Journal*, vol. 3, p. 254, 2010.

[6] Yogendra S. Dohare, Tanmoy Maity, P. S. Das, and P. S. Paul, "Wireless communication and environment monitoring in underground coal mines – Review," *IETE Tech. Rev.*, vol. 32, no. 2, pp. 140–150, 2015.

[7] Serhan Yarkan, Sabih G¨uzelg¨oz, H¨useyin Arslan, and Robin R. Murphy, "Underground mine communications: A survey," *IEEE Communication Survey & Tutorials*, vol. 11, no. 3, Third quarter 2009.

[8] A. M. Donoghue. "Occupational health hazards in mining: An overview." *Occup. Med. (Lond)*, vol. 54, no. 5, pp. 283–289, 2004, doi: 10.1093/occmed/kqh072, PMID: 15289583.

[9] R. Agnelo, C. V. Silva, and C. Mehmet, "Communication with above devices in wireless underground sensor networks: An empirical study," in *Communications (ICC). IEEE International Conference Proceedings*, IEEE Commun. Soc., Cape Town, pp. 23–27, 2010.

[10] O. Green, E. S. Nadimi, V. Blanes, R. N. Jorgensen, and C. G. Sorensen CG, "Monitoring and modeling temperature variations inside silage stacks using novel wireless sensor networks," *Comput. Electron Agric.*, vol. 69, no. 1, pp. 149–157, 2009.

[11] X. Q. Yu, Z. L. Zhang, and W. T. Han, "Evaluation of communication in wireless underground sensor networks," in *IOP Conference*

Series: Earth and Environmental Science, vol. 69, 3rd International Conference on Advances in Energy, Environment and Chemical Engineering, Chengdu, China, 2017.

[12] X. Dong, M. C. Vuran, and S. Irmak, "Autonomous precision agriculture through integration of wireless underground sensor networks with center pivot irrigation systems," *Ad Hoc Networks Journal*, vol. 11, pp. 1975–1987, 2013.

[13] X. Zhang, A. Andreyev, C. Zumpf, M. C. Negri, S. Guha, and M. Ghosh, "Thoreau: A subterranean wireless sensing network for agriculture and the environment," in *IEEE Conference on Computer Communications Workshops (INFOCOM WKSHPS)*, pp. 78–84, 2017.

[14] Coen J. Ritsema, Henk Kuipers, and Leon Kleiboer, "A new wireless underground network system for continuous monitoring of soil water contents," *Water Resources Research*, vol. 45, 2010, doi: 10.1029/2008W

[15] I. F. Akyildiz and E. P. Stuntebeck, "Wireless underground sensor networks: Research challenges," *Ad Hoc Networks Journal*, vol. 4, pp. 669–686, 2006.

[16] M. C. Vuran and A. R. Silva, "Communication through Soil in Wireless Underground Sensor Networks Theory and Practice, *Sensor Networks: where Theory Meets Practice*. Springer, 2009.

[17] A. R. Silva and M. C. Vuran, "Development of a Testbed for Wireless Underground Sensor Networks," *EURASIP Journal on Wireless Communications and Networking*, vol. 2010, p. 620307, 2010.

[18] X. Dong, M. C. Vuran, and S. Irmak, "Autonomous precision agriculture through integration of wireless underground sensor networks with center pivot irrigation systems," *Ad Hoc Networks Journal*, vol. 11, pp. 1975–1987, 2013.

[19] A. Markham, N. Trigoni, D. W. Macdonald, and S. A. Ellwood, "Underground localization in 3-D using magneto-inductive tracking," *IEEE Sensors Journal*, vol. 12, pp. 1809–1816, 2012.

[20] X. Dong and M. C. Vuran, "Spatio-temporal soil moisture measurement with wireless underground sensor networks," in Ad Hoc Networking Workshop, 2010. Med-Hoc-Net 2010. 9th IFIP Annual Mediterranean, Juan-les-pins, France, 2010.

[21] C. J. Ritsema, H. Kuipers, L. Kleiboer, E. Elsen, K. Oostindie, J.G. Wesseling, J. Wolthuis, and P. Havinga, "A new wireless underground network system for continuous monitoring of soil water contents," *Water Resources Research Journal*, vol. 45, pp. 1–9, 2009.

[22] I. F. Akyildiz, Z. Sun, and M. C. Vuran, "Signal propagation techniques for wireless underground communication networks," *Physical Communication Journal (Elsevier)*, vol. 2, pp. 167–183, 2009.

[23] X. Dong and M. C. Vuran, "A channel model for wireless underground sensor networks using lateral waves," in *IEEE Globecom 2011*, Houston, TX, USA, 2011.

14

PREDICTING EFFECTIVENESS OF SOLAR POND HEAT EXCHANGER WITH LTES CONTAINING CuO NANOPARTICLE USING MACHINE LEARNING

K. KARUNAMURTHY[1], G. SUGANYA[1], M. ANANTHI[2], AND T. SUBHA[2]

[1]Associate Professor, Vellore Institute of Technology Chennai, Tamil Nadu, India
[2]Sri Sairam Engineering College, Chennai, Tamil Nadu, India

Contents

DOI: 10.1201/9781003119838-14

Introduction

Solar energy is clean, free, and available abundantly [1]. It is the primary duty of humanity to be socially responsible to adopt the sustainable development goals (SDGs) of the United Nations, and this is possible if mankind utilizes this renewable source of energy for the progress of society. The solar radiation striking our planet is 174 PW (174 × 1012 kW) at its upper atmosphere and the average amount of solar energy absorbed by the landmasses, oceans, and atmosphere is approximately 121.8 × 1012 kW. The rate at which solar energy is received outside the earth's atmosphere perpendicular to the solar beam at the earth's mean distance from the sun is 1.373 kW/m^2.

Solar energy is utilized by two methods, such as solar photovoltaic (SPV) and solar thermal technology (STT) [2]; the applications of STT has a broader range than SPV. STT is used to collect and store solar energy in the form of (i) low-temperature heat (< 100°C) or (ii) high-temperature heat (up to 500°C). The various solar collectors used are flat plate collectors (FPC), concentrated solar thermal collectors (CSCs), and evacuated tube collectors (ETCs). FPC and ETCs are used for low-temperature applications like domestic hot water supply, space heating of the residential, commercial and institutional building, and for industrial process heat. CSCs provide comparatively high-grade thermal energy, which can even operate heat engines. A solar pond is a type of collecting and storing large amounts of low-temperature heat with less expense in a salt gradient pool of water. Solar ponds provide thermal energy for longer periods in the temperature range of 70°C to 100°C.

Solar Pond

Solar ponds (SPs) are water bodies with varying salinity gradient called halocline, i.e., the salinity of the pond increases with depth. These solar ponds possess three different distinct zones viz., upper convective zone (UCZ), non-convective zone (NCZ), and lower convective zone (LCZ). These zones have different densities due to the presence of salinity, and they store the thermal energy by arresting the convective effects. The solar radiation reaching the bottom of the pond heats the water of the LCZ as the density of the LCZ is higher, the bulk motion of water to the top surface of the pond is prevented, thereby the thermal energy is trapped within the LCZ itself. This collected and stored thermal energy at the LCZ is used for various engineering applications. Thus solar ponds are one of the most effective ways of capturing and storing solar energy [3]. The applications of solar ponds include industrial process heating, desalination, greenhouse heating, crop drying, electric power generation, refrigeration, and air conditioning. A typical schematic diagram of a solar pond is represented in Figure 14.1.

The stored energy of the solar pond is used for the above-said various applications by incorporating internal or external heat exchangers. The main concern with this low-temperature energy storage system (LTES) is its poor thermal efficiency of 15% to 25% and the period of operation of the solar pond. The thermal efficiency of a solar pond is defined as the ratio of total energy extracted from the solar pond with the help of the heat exchanger to the total solar radiation falling on the surface of the solar pond. There are different

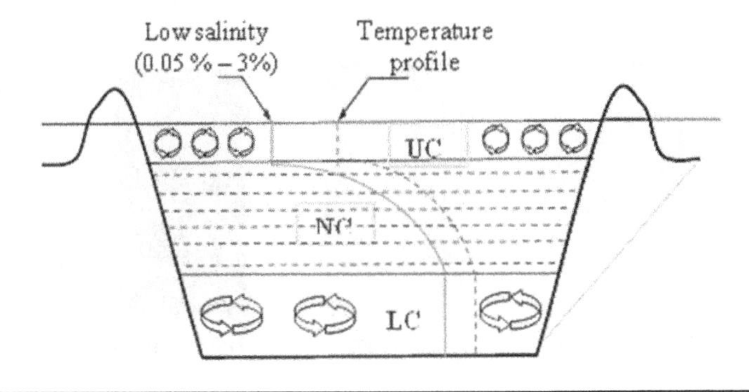

Figure 14.1 Typical Sketch of a Solar Pond

works reported by researchers in the past literature. The effect of insulation thickness of the side wall on the SP efficiency was studied [4]. The current research work carried out is to increase the performance of the solar pond by providing twisted tapes in the flow passage of the heat exchanger, and by placing encapsulated PCM blended with CuO nanoparticles in the LCZ as an LTES system to increase the period of operation.

Low-Temperature Energy Storage System (LTES) for Solar Pond

There is a great potential in various energy-related applications for phase change material (PCM)-based thermal energy storage (TES) [5]. The availability of solar energy is limited to the day, and hence the solar pond can operate only during the day time. Thus LTES is mandatory to operate the solar pond in the absence of solar radiation and to increase the period of operation in a day. LTES tries to match the demand and supply of the solar energy. The two ways by which thermal energy can be stored are sensible heat storage and latent heat storage. However, latent heat energy storage is preferred as the energy density of storage is more. In this research, the LTES is coupled with the LCZ of the solar pond; this is achieved by introducing encapsulating PCM in copper capsules. There are wide varieties of PCM available; however, in this experimentation, paraffin was selected as the PCM, as it is readily available in the market and also the physical and chemical properties of paraffin do not change even after thousands of charging and discharging cycle. The three processes in the thermal cycle of LTES are (i) charging, (ii) storing, and (iii) discharging. Figure 14.2 illustrates these three processes.

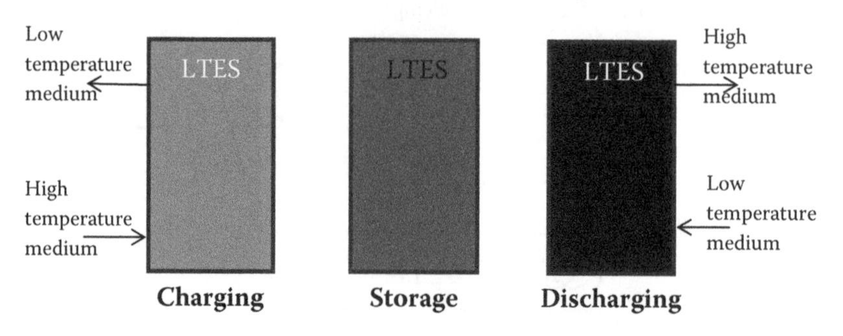

Figure 14.2 Processes Involved in LTES

Background Study

Usage of PCM to increase thermal conductivity is discussed by many researchers. Shu-Rong et al. proposed the method of blending multi- walled Carbon nanotubes in different proportions with paraffin to increase the thermal conductivity [6]. In this research, CuO nanoparticles are added with to the paraffin. The increase in the rate of convective heat transfer by using swirl tapes is analyzed through a numerical study for 3-D heat transfer and flow behaviors in circular tubes with loose-fit multiple channel [7]. Karunamurthy et al. discussed the effect of Al_2O_3 nanoparticles to increase conductivity [8].

Machine learning (ML) techniques are profoundly used to reduce a number of experiments conducted physically thereby reducing budget, manpower, and natural resources. Use of linear regression for predicting output values where the values are continuous with greater correlation between predictor and response variables is discussed by researchers. Since the amount of data collected is relatively low, researchers suggest the use of linear regression for prediction [9,10]. Christopher et al. compared the advantages of using linear regression with neural networks and vice versa [11]. Other applications of the method are discussed by Mahesh Gadhavi et al. [12]. A relative study of various kinds of applications using regression strengthens the selection of model for the proposed study.

Experimental Setup

The experimental setup fabricated is as per Huseyin Kurt et al. (2006) and it is presented in Figures 14.3 and 14.4.

Laboratory model solar pond of three numbers of dimension 50 cm × 60 cm × 60 cm is fabricated using 0.15-cm galvanized iron sheet. A heat exchanger made up of a copper tube of 1.2 cm outer diameter and 1.05 cm inner diameter is placed in the LCZ of the solar ponds with five passes. The three different solar ponds used for experimentation are (i) ordinary SP, (ii) SP with LTES (PCM blended CuO nanoparticle), and (iii) SP with LTES (PCM blended CuO nanoparticle) and swirl tape. Experiments are conducted and

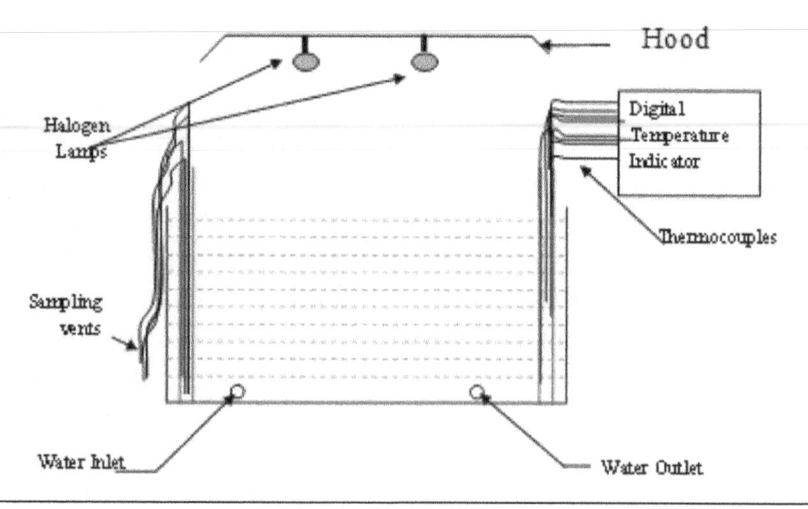

Figure 14.3 Solar Pond Experimental Setup

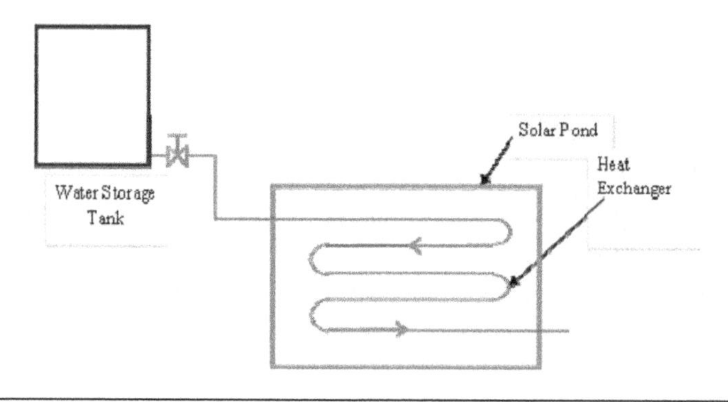

Figure 14.4 Schematic Sketch of Heat Exchanger in LCZ

swirl tape of optimum twist ratio of 6.36 is placed in the flow passage to augment the rate of heat transfer. Swirl tapes provide secondary flow and turbulence, which enhances the rate of heat transfer.

The PCM paraffin (n-Tricosane C23H48) of shell make is chosen for thermal energy storage and the same is encapsulated and placed in the LCZ of the solar pond as LTES for experimentation. However, the thermal conductivity of paraffin PCM is 0.214 W/mK (solid) and 0.15 W/mK (liquid). This is poor and to increase the thermal conductivity, CuO nanoparticles are added.

There is a limit for the percentage of CuO nanoparticles that can be blended to the PCM because the addition of too many CuO nanoparticles with PCM will also lead to agglomeration which is not

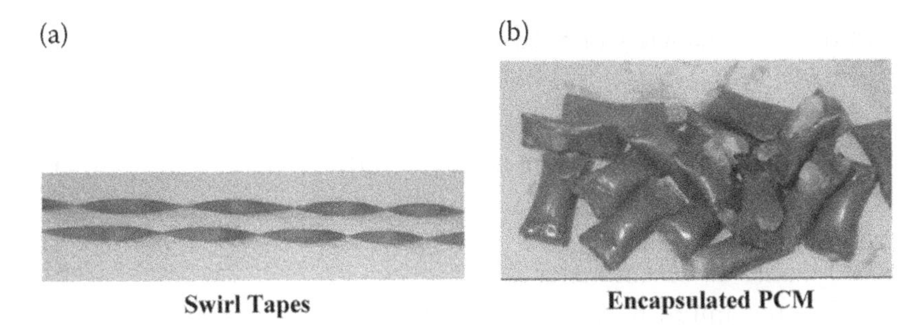

Figure 14.5 (a) Swirl Tapes and (b) Encapsulated PCM

preferred. In this aspect, experiments are conducted to identify the optimum proportion of CuO nanoparticles that can be added to the PCM. Encapsulated PCM blended with CuO nanoparticles and swirl tapes used are represented in Figure 14.5(a) and (b).

Experimentation

Experiments were conducted on all the three solar ponds under solar irradiated condition. Heat transfer fluid used is water, which is allowed to flow through the flow passage of the heat exchanger under laminar and turbulent flow conditions with Reynolds's number of 1804 and 9020, respectively. The objective of the study is to have higher outlet temperature of the heat exchanger, indicating better performance. The various parameters related to the outlet water temperature are ambient temperature, water inlet temperature, solar irradiation, flow rate, and LCZ temperature.

The temperatures such as ambient, inlet, outlet, LCZ, NCZ, and UCZ were measured using "K"-type thermocouple connected to digital temperature indicator. The solar radiation reaching the solar pond is measured using pyranometer, and the salinity of the LCZ, NCZ, and UCZ is measured daily using pyranometer to ensure the existence of halocline. Sampling vents were provided to get the sample from LCZ, NCZ, and UCZ. The readings are recorded every 30 minutes in a day from 8:00 AM to 6:00 PM over a period of 41 days during May–June 2019. The average values are tabulated for analyzing the solar pond performance parameters, such as outlet water temperature, temperature rise, rate of heat transfer, efficiency of solar pond, and effectiveness of heat exchanger.

Performance Parameters of Solar Pond

i. Rate of heat transfer

$$Q = m\,cp(To - Ti), \tag{14.1}$$

where
Q – rate of heat transfer (W)
Q' – maximum heat transfer (W)
$\dot{m}$ – mass flow rate (kg/s)
Cp – specific heat capacity (J/kg °C)
To – outlet temperature (°C)
Ti– inlet temperature (°C)
TLCZ – temperature of the LCZ (°C)

ii. Maximum rate of heat transfer (Q') $Q' = \dot{m}Cp(TLCZ-Ti)$

iii. Effectiveness of heat exchanger

$$\in = \frac{Q}{Q'} \tag{14.2}$$

iv. Efficiency of solar pond

$$\eta = \frac{\text{Useful heat carried away by flow medium}}{\text{Radiation incident} * \text{Surface area of solar pond}}$$

Modeling Using Machine Learning

Intelligence, the need for the era, is expected not only from human, but also from machines. ML is the field of study that gives computers the competence to be intelligent without being explicitly programmed. Applications of ML range from day-to-day application like automatic switching on/off of household motors to more complex applications like controlling satellite operations.

In course of this, ML finds an integral part in Industry 4.0 transforming applications in mechanical engineering domain to be more intelligent using techniques like predictive maintenance, demand prediction, quality control, etc. Generally, numerous experiments would be carried out on solar pond and the input parameters supplied and output obtained will be stored as a relational dataset during experimentation. This collected dataset will be analyzed to identify input values that can end up in optimal performance. But, this process needs exhaustive repetition of experiments and will lead to waste of materials, money, time and human resources. Also, the feasibility of witnessing the optimality of the solution cannot be granted.

ML techniques can be a boon in such cases where the machine can learn the pattern that exists in the dataset resulted from limited experimentation. The use of ML techniques for predicting performance parameters is discussed by researchers [9]. Once patterns are understood through proper modeling, any number of predictions is possible without physical experimentation. ML algorithms witness to reduce wastage of resources, budget, and time. Figure 14.6 represents the process of modeling solar pond performance prediction using multi-linear regression.

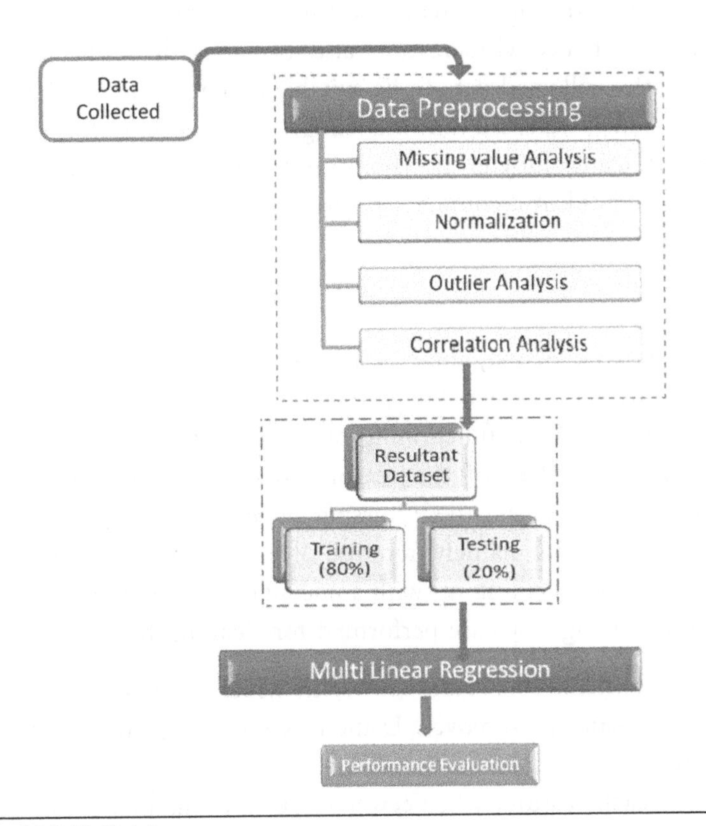

Figure 14.6 ML Modeling

Data Collection and Preprocessing

The collected data are stored in a Microsoft excel document where the columns represent the features and rows represent data for every experimentation. Around 820 recordings are created over a period of 41 days. Ambient temperature, water inlet temperature, solar radiation flux, and LCZ temperature are considered to be input parameters and outlet water temperature is recorded as output parameter. LCZ and water temperatures are recorded for all three ponds.

Choice of right features and reliable and complete data are most important for accurate prediction of output. Measuring instruments such as Pyranometer, thermocouples and Pyranometer used for recording the values are made of electronic components and hence are prone to errors. Also, data are taken at different time periods and the possible environmental changes may also affect recording of data. The format and range of data also vary from one feature to another feature and hence visualization also becomes difficult. This step prepares the collected data to be suitable and accurate for the inputs of the models selected for modeling. The entire process is programmed using Jupyter notebook and the data are stored using Microsoft Excel in a comma separated values (CSV) format.

Data Cleaning

During data collection, there is a possibility for some values to be missed. This may happen due to non-availability of data or some data validation rule or malfunctioning of some device or due to human error. But regardless of the reason, the dataset should be made complete by processing these missing values. Several approaches are used by researchers, namely, removal of missing data, filling up with some static value, filling up with some statistical data, etc.

The following steps are performed for cleaning the data:

- Experimentation values (rows) in which three or more values are missing are removed. Using this approach, a total of 18 rows were removed.
- For each feature, when fewer values are missing, mean is calculated and is used to fill the missing values. Figure 14.7

```
df.isnull().sum()

Ambient Temperature             1
Inlet Water Temperature         1
Incident Radiation Flux         3
LCZ temperature                 1
Outlet Water Temperature        4
 LCZ1 temperature               0
Outlet water temperature 1      1
 LCZ temperature 2              1
Outlet water temperature 2      1
dtype: int64
```

```
df.isnull().sum()

Ambient Temperature             0
Inlet Water Temperature         0
Incident Radiation Flux         0
LCZ temperature                 0
Outlet Water Temperature        0
 LCZ1 temperature               0
Outlet water temperature 1      0
 LCZ temperature 2              0
Outlet water temperature 2      0
dtype: int64
```

Figure 14.7 Missing Values (before and after Filling Up).

represents the status of missing values in the original dataset and after filling up with mean.

Normalization

Data for each attribute are collected in different ranges and units. ML modeling algorithms work on numeric data and hence variables are measured at different scales. Temperatures are measured in degree Celsius and Incident solar radiation flux is measured in Watt/m^2. The range of values corresponding to various features does not contribute in the identical way to the machine model fitting and might end up creating a bias to the overall system. Also, visualization becomes very difficult with varying ranges of data. Hence, normalization is applied to convert ranges of all data between 0 and 1. Min-Max normalization, as in Equation 14.3, is used in our modeling.

$$x_{norm} = \frac{x - \min(x)}{\max(x) - \min(x)} \tag{14.3}$$

Outlier Analysis

During data collection, there might be situations in which the electronic components may malfunction resulting in random data. Though recorded by human being, such randomness may go unnoticed leading to improper training. Hence outliers, the observations that lie at an unusual distance from other values in the collected data are to be

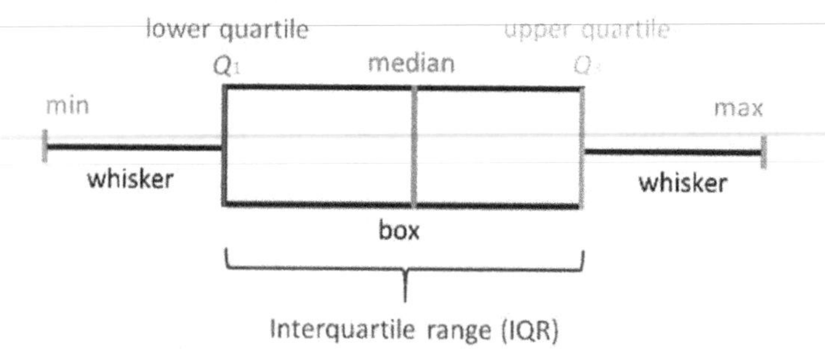

Figure 14.8 Box Plot for Outlier Analysis

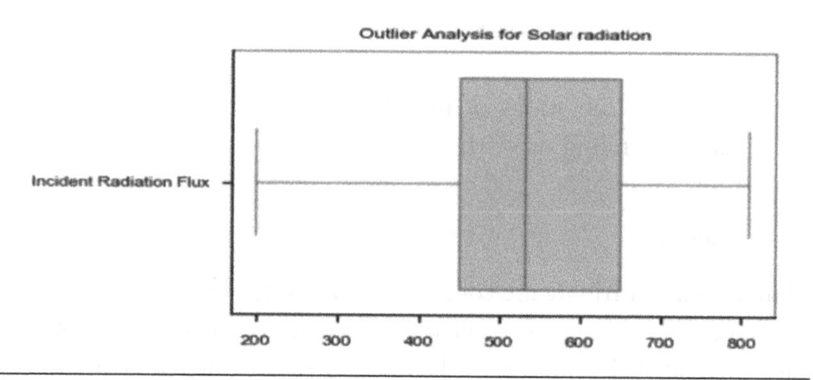

Figure 14.9 Outlier Analysis for "Solar Radiation Flux"

identified and removed. Box plots are used to perform this analysis where the data that lie outside of 1.5 * interquartile range on both the ends are considered as outliers and are removed. Figure 14.8 represents the various ranges identified using box plot and Figure 14.9 represents outlier analysis for inlet solar radiation flux.

Correlation Analysis

Since the values of output attributes is continuous that is spread over a range, regression modeling is required. Linear regression (LR) is the most simple and commonly used technique for regression [9]. Logically, LR models the relationship between the input and output variables by fitting a straight line with the following assumptions:

1. Recordings are independent among them.
2. Relationship between input and output remains linear.
3. Very little or no correlation among the input attributes.

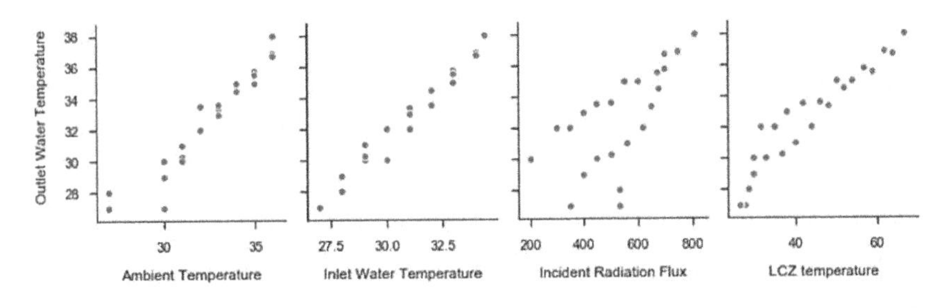

Figure 14.10 Scatter Plots for Correlation Analysis

Assumption 1 is satisfied since the experimentation is done at different timings and the relationship is not considered. Assumptions 2 and 3 are observed using correlation analysis, a method that is used to calculate the strength of relationship between different features. Scatter plots are used to visualize the correlation between features graphically. The correlation between independent features to output feature is visualized using scatter plot and is presented in Figure 14.10.

To analyze the numeric distribution of data, heat map is a better graphical representation tool in which data values are represented as colors. Variations from white to black indicate the distribution of range of data. Correlation values can be visually understood using heat map. Figure 14.11 presents a heat map representing the correlation

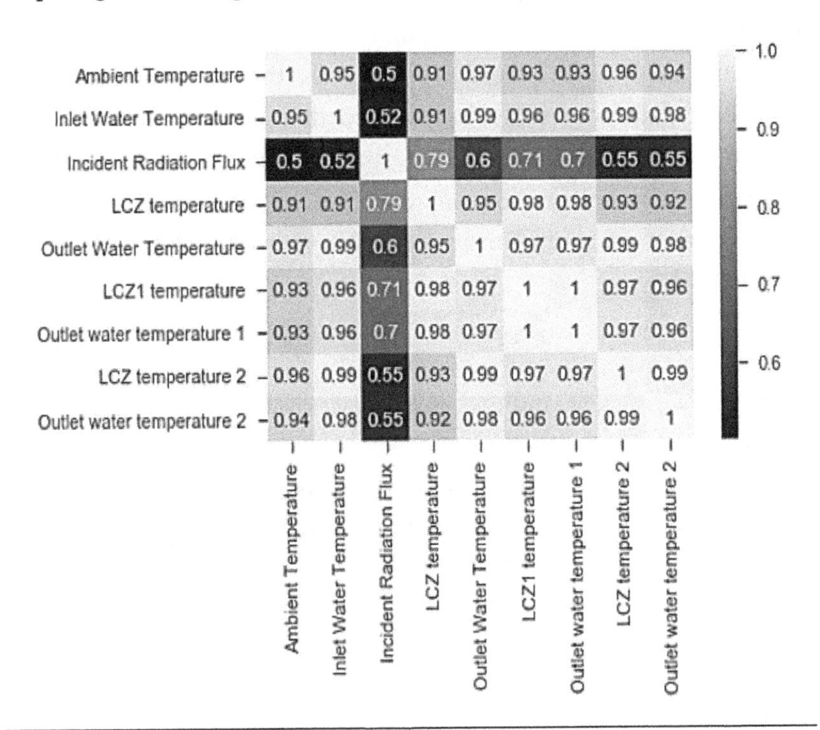

Figure 14.11 Heat Map Representing Correlation Values

between the features. It is evident from the figure that all the outlet temperatures are positively correlated with maximum correlation with all input variables and hence linear regression is chosen for modeling the dataset.

Multi-Linear Regression

Predictive modeling is very useful for forecasting future conclusions and assessing metrics that are impractical or difficult causing wastages in physical measure. When the output data are continuously varying in a finite range, regression models are the best choice. Many regressive models are proposed by research community that varies in understanding differing patterns in data. Linear regression is used for predicting the values of output variables in situations where training data are less [10].

Linear regression is a basic supervised regressive method that arrives at a straight line with constant slope. It's used to predict values within a continuous range, (e.g., sales, price) rather than trying to classify them into categories. Since, the number of independent variables is more than one, multi-linear regression is chosen for this task. Regression is an eager learning method that abstracts the data during the training itself. Regression method returns a set of straight line to a graph such that the line passes through the maximum number of data points of the collected dataset. This line fitted is called the hypothesis and is used to predict the dependent value for any given set of independent values $(x_1, x_2, \ldots, x_n)$.

The output of multi-linear regression model is represented by Equation 14.4.

$$y_{pred} = m_1 x_1 + m_2 x_2 + \ldots + m_n x_n + m_0 + c \tag{14.4}$$

where, $m_1, m_2, \ldots, m_n$ indicate the impact of each attribute over the output.

$x_1, x_2, \ldots, x_n$ indicate the values of independent variables.

m_0 refers to the common bias that exists in the dataset.

c represents the error of the model.

y_{pred} refers to the predicted output.

The coefficients m_1, m_2, ..., m_n measure the effect of each independent variable in association with the effects of all the other independent variables in the model. Thus, these coefficients measure the marginal effects of the independent variables. The difference between the actual and predicted output is termed as the error or deviation. The average of all errors in the dataset is termed as mean error of the dataset caused by the model. This error is to be reduced for the model to be more accurate and be reliable for the given dataset. The least squares method offers a way of choosing the coefficients properly by minimizing the sum of the squared errors. Least squares plot is used to check the distribution of data with respect to an approximate line fitting. Figure 14.12 depicts the least squares plot for various input parameters to output parameter.

The accuracy of the model is evaluated using various metrics like mean absolute error (MAE), mean square error (MSE), or root mean square method (RMSE). We have used RMSE as the standard error analysis metric. Equation 14.5 defines the RMSE calculation.

$$\text{Root Mean Square Error} = \sqrt{\frac{\sum_{i=1}^{n}(x^i - x'^i)^2}{n}} \tag{14.5}$$

where x_i represents the actual value of input attribute.

x'_i indicates the predicted value of dependent variable(s).

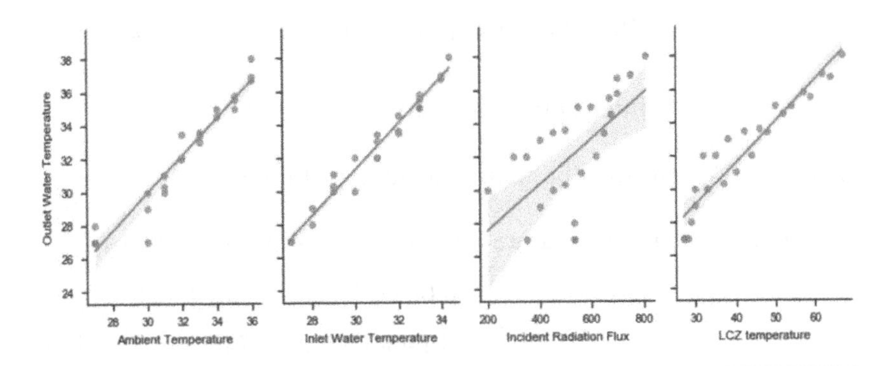

Figure 14.12 Least Squares Line Plot

Splitting Dataset into Training and Testing Data

The dataset is divided into two random sets of data, namely, training and testing using 80 to 20 proportion. The first set comprising 80% of the data termed as training dataset is used to fit the model. The training set covers the set of inputs with known output that the model uses to learn to generalize other data. The second set comprising 20% of the data, termed as testing dataset is used to check whether the fitting is accurate and acceptable. The package train-test-split in the scikit-learn library of Python is used to divide the dataset. This package takes care of random splitting of rows to avoid bias that may be caused due to environmental conditions.

The dataset is divided into input and output features termed as "X" and "Y," respectively. Using the train-test-split function of scikit-learn library, X and Y are divided into four subsets, namely, X_train, X_test, Y_train, and Y_test. The dimensions of these are (642,4), (642,), (160,4), and (160,). Three such datasets representing three different solar ponds were taken for modeling.

Fitting and Prediction

Finding the best estimates of the coefficients by applying a specific model is called "fitting" the model. This is performed using the training dataset and the process is termed as learning or training the model. "linear_model" of scikit-learn library is used to train the training data using multi-linear regression. Once learnt or trained with the patterns of the data, the model returns a set of four coefficients.

$$\text{Coefficients: } [0.46025584 \;\; 1.04712856 \;\; 0.00551125 - 0.07416891]$$

$$\text{Intercept: } -14.436684821577494$$

The equation can then be written as in Equation 14.6.

$$\begin{aligned} \text{Outlet water temperature} = \; & 0.46025584_1 x_1 + 1.04712856 x_2 \\ & + 0.00551125 x_3 - 0.07416891 x_4 \\ & - 14.436684821577494 \end{aligned} \quad (14.6)$$

```
y_predicted = regress.predict(train_x)
print("Actual","        ","Predicted")
for i in range(0,len(train_x)):
    print(train_y[i],"     ",y_predicted[i])

Actual          Predicted
27.0        27.56983681932116
29.0        28.670020938702486
30.0        29.770205058083818
30.3        30.20934754108943
31.0        30.31751555992593
32.0        31.858899062250877
33.4        33.23494518641407
34.5        34.58343508088364
35.8        35.85775606285752
```

Figure 14.13 Sample Analysis of Actual vs. Predicted Value

The outlet water temperature for the test data is then predicted using the "predict" function of the scikit-learn library. The predicted value of testing data along with original output value in the dataset is compared. Figure 14.13 depicts a sample comparison.

Results and Discussion

Accuracy

The accuracy is used as an important measure to find out the correctness of prediction [11]. It is defined to be the ratio between numbers of correct predictions to total number of predictions. Accuracy of the prediction is tested visually using a 3-D graph that enables visualization with more than two features. Figure 14.14 depicts original values of outlet water temperature and predicted values of predictor variables, outlet water temperature with ambient and inlet water temperature.

MAE, MSE, and RMSE are calculated and are given below. The values are found to be less than the assumed threshold and are satisfactory.

Goodness of Fit

To understand the goodness of fit with respect to variations in data, R^2 value, coefficient of determination is calculated. This value reflects

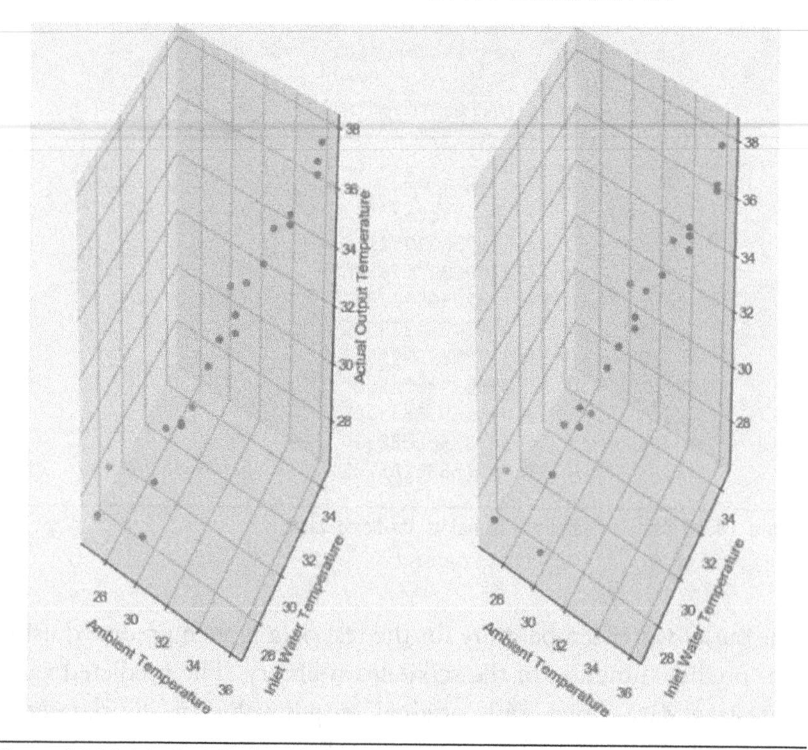

Figure 14.14 Actual vs. Predicted Values

the proportion of variation in the independent variables as defined by the regression model. If the predictions are close to the actual values, the values of R^2 will be close to 1. On the other hand, if the predictions are unrelated to the actual values, then R^2 will be close to 0. The value obtained for the proposed model is 0.9928512303131685 and hence is close to 1.

Future Works

Solar ponds are still in amateur stage and research is to be carried out for augmenting its performance and to commercialize. The materials for thermal energy storage at the LCZ, non-corrosive (materials) layers for the side wall, and the bottom of the pond have lot of scope. This is a low-temperature energy storage method and needs optimization of process parameters in order to commercialize for generating electric power. There are many process parameters involved with the operation of solar pond. The above work uses only few parameters and handled dataset taken only for a certain period of time. Also, the collinearity between the predictor variables is not properly realized by our model.

Future works may concentrate on handling large volume of data with proper analyzing on relationship between the independent variables. Also, metrics such as accuracy derived from ML modeling will not be useful for adding business value to the stated problem by researchers. The optimal values of input parameters can be obtained from applying optimization techniques.

References

[1] The Solar Vision: A 100-day Agenda of Solar Energy Industries Association. https://www.seia.org/initiatives/about-solar-energy.

[2] https://energypedia.info/wiki/Portal:Solar.

[3] Abhishek Kumar, and Ranjan Das, "Effect of peripheral heat conduction in salt-gradient solar ponds," *Journal of Energy Storage*, November 2020. 10.1016/j.est.2020.102084.

[4] Assad H. Sayer, Hazim Al-Hussainia, and Alasdair N. Campbell, "New theoretical modelling of heat transfer in solar ponds," *Solar Energy*, vol. 125, pp. 207–218, 2016. 10.1016/j.solener.2015.12.015.

[5] Binjian Nie, Anabel Palacios, Boyang Zou, Jiaxu Liu, Tongtong Zhang, and Yunren Li, "Review on phase change materials for cold thermal energy storage applications," *Renewable and Sustainable Energy Reviews*, vol. 134, 2020. 10.1016/j.rser.2020.110340.

[6] Shu-Rong Yan, Rasool Kalbasi, Aliakbar Karimipour, and Masoud Afrand, "Improving the thermal conductivity of paraffin by incorporating MWCNTs nanoparticles," *Journal of Thermal Analysis and Calorimetry*, 2020. 10.1007/s10973-020-09819-0.

[7] Suttisak Kunlabud, Varesa Chuwattanakul, Vichan Kongkaitpaiboon, Pitak Promthaisong, and Smith Eiamsa-ard, "Heat transfer in turbulent tube flow inserted with loose-fit multi-channel twisted tapes as swirl generators," *Theoretical and Applied Mechanics Letters*, vol. 7, no. 6, pp. 372–378, 2017. 10.1016/j.taml.2017.11.011.

[8] Karunamurthy, K., Rajesh, M. R., Vijaypal, B., and Kumar, A., "Thermal conductivity and charging & discharging characteristics of a thermal energy storage system blended with Al2O3 nanoparticles," *Nano Hybrids and Composites*, vol. 17, pp. 10–17. 2017. 10.4028/www.scientific.net/nhc.17.10.

[9] D. C. Montgomery, E. A. Peck, and G. G. Vining, *Introduction to linear regression analysis*, John Wiley & Sons, 2015.

[10] K. Karunamurthy, R. Manimaran, and M. Chandrasekar, "Prediction of solar pond performance parameters using artificial neural network," *International Journal of Computer Aided Engineering and Technology*, vol. 11, no. 2, pp. 141–150, 2019. 10.1016/j.taml.2017.11.011
[11] Christopher Osita Anyaeche, and Desmond Eseoghene Ighravwe, "Predicting performance measures using linear regression and neural network: A comparison," *African Journal of Engineering Research*, vol. 1, no. 3, pp. 84–89, 2013.
[12] Mahesh Gadhavi, and Chirag Patel, "Student Final Grade Prediction Basedon Linear Regression," *Indian Journal of Computer Science and Engineering*, vol. 8, no. 3, pp. 274–279, 2017.

Index

Note: *Italicized* page numbers refer to figures, **bold** page numbers refer to tables